To an inspirational
friend and teacher.

Best wishes,

Steve

Energy Strategies:
Toward a Solar Future

Energy Strategies: Toward a Solar Future

A Report of
The Union of Concerned Scientists

EDITORS

Henry W. Kendall and Steven J. Nadis

CONTRIBUTORS

Daniel F. Ford   Henry W. Kendall
Carl J. Hocevar   Ronnie D. Lipschutz
David J. Jhirad   Steven J. Nadis

Ballinger Publishing Company ● Cambridge, Massachusetts
A Subsidiary of Harper & Row, Publishers, Inc.

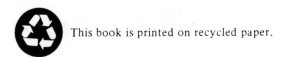
International Standard Book Number: ISBN 0-88410-622-5

Library of Congress Catalog Card Number: 79-23757

Printed in the United States of America

Library of Congress Cataloging in Publication Data

Kendall, Henry Way, 1926-
  Energy strategies.

  Includes index.
  1. Power resources—United States. 2. Solar energy—United States. 3. Energy policy—United States. I. Nadis, Steven J., joint author. II. Union of concerned Scientists. III. Title.
HD9502. U52K46   333.7   79-23757
ISBN 0-88410-622-5

Contents

List of Figures

List of Maps

List of Photographs

List of Tables

Preface

Despite the obvious dependence of the nation's material prosperity on an adequate supply of energy, the United States has never had a coherent, long-term strategy for assuring reliable fuel supplies and for developing satisfactory energy production, conversion, transmission, and storage technologies. National "policy" governing the management of energy resources, energy pricing, energy facility regulation, and research on new energy technologies has been capriciously and unsystematically formulated. Successive administrations have generally limited their energy decisionmaking to short-term crisis management, have carried out few competent technical appraisals of the critical energy policy issues, and have failed to bring about a national consensus on what needs to be done to assure the vital energy supplies required by the country.

The decades of negligence in the formulation of a thoughtful national energy policy have already had a number of adverse consequences and pose future difficulties for the United States. Among the many far-reaching problems that have been created are:

- A national level of energy use that by any reasonable standard is excessive, wasteful, and inefficient;
- The consequent impending exhaustion of scarce domestic supplies of oil and natural gas;
- The consequent dependence on imported fuels, which has now led to a complex of national security problems and damaging balance of payments deficits;

- Conventional fossil fuel technology that has had painful environmental and public health repercussions and that poses long-term threats to the global climate;
- Haphazard development of new energy sources and technologies with little or no priority assigned to the most promising ones;
- An excessive commitment to nuclear power technology whose economics, reliability, safety, and national security effects are highly uncertain.

To the extent that the federal government has had any forward-looking energy strategy, this has been principally reflected in a narrowly focused and poorly managed effort to exploit nuclear energy for commercial power production. Since World War II, and up until the last few years, virtually all federal energy research and development moneys were devoted to nuclear technology. These expenditures were accompanied by official projections that called for a massive nuclear power plant construction program and slated nuclear power to become the dominant source of the nation's electricity by the year 2000. The nuclear program is currently in a state of remission, if not collapse, owing to downward revision of electric power growth projections, uncontrolled nuclear plant capital cost increases, and an extensive public controversy over the possible side effects of nuclear power—nuclear radiation accidents, problems posed by long-lived radioactive waste materials, and unintended nuclear weapons proliferation engendered by large civilian nuclear power programs.

The Union of Concerned Scientists has been heavily involved since 1971 in research on the basic problems connected with nuclear power. Because of the unresolved technical problems, together with its doubtful economics, it became apparent that nuclear power may be unable to make the large contribution that the federal government had projected. Accordingly, the Union of Concerned Scientists initiated a systematic study effort to assess the alternatives to nuclear power and to answer the question, If not nuclear power, what? The results of this UCS research effort are set forth in this report.

The UCS Energy Study examined in great detail the individual components of the present U.S. energy system and, in terms of explicit criteria, set out to put together a new energy system that would supply energy in the quantities and forms required by the U.S. economy. Rather than simply extrapolating growth rates for the consumption of the various forms of energy, as has been done in conventional analyses by the federal government, electric utility companies, and major oil suppliers, the UCS Energy Study looked sector by sector at energy use in the U.S. economy, at applicable techniques for improving energy efficiency in those sectors, and at the energy requirements for various levels of improvement in the nation's overall standard of living. Having established the range of future energy requirements in the major sectors of the U.S. economy, the study then turned to an assessment of the energy resources and technologies that can be used to meet these needs. The energy technology assessments looked at resource availability, technical feasibility, economics, and environmental effects and, most importantly, tried to apply an even-handed analysis to all of the various technological options.

It is particularly important that a balanced and dispassionate view be taken of the so-called "alternate energy sources"—such as geothermal power, tidal power, and solar energy in its various forms including wind energy, hydroelectric power, biomass, and ocean thermal gradients—that are now the focus of great interest. Acute concern about nuclear power has led to strong advocacy for these various nonconventional energy sources. Our research concluded that some of the sanguine assessments that are being made of alternate energy sources and their potential contribution to U.S. energy supplies have a thin technical foundation; many of the claims are based more on the type of contagious enthusiasm that once surrounded nuclear power than on the results of an objective assessment of the technical and economic feasibility of these options. Our studies of nuclear power have shown how program commitments made on the basis of uncritically accepted and highly optimistic technical premises have led to many of the subsequent difficulties of this technology. It would be a grave mistake to repeat this style of analysis in a selection of future energy technologies. Systematic, critical, and independent technical assessments must be applied to the alternate energy sources. When this is done, as the results of the UCS study show, many of these alternate energy sources are capable, at best, of only limited usefulness and hardly stand as a panacea to the nation's energy problems. Yet on the positive side, a critical review of the alternate technologies helps identify those with the greatest likelihood of making a substantial contribution, information that can be used profitably in establishing research priorities. Indeed, the overall assessment that the UCS Energy Study offers is that shrewdly selected applications of solar energy have very favorable prospects for becoming the basic supplier of the nation's long-term energy requirements.

Acknowledgments

Important contributions to the UCS Energy Study have been made by a number of UCS staff members and by others who gave voluntarily of their time and expertise. We are especially grateful for the generous help provided by Bruno Coppi, Peter Franchot, Richard Graves, David Inglis, James Mac-Kenzie, Donald Scroggins, Robert Toy, Eric Van Loon, and Frank von Hippel. The editors, of course, accept full responsibility for the final content and conclusions of this report. Thanks should also be extended to Janice Candelora, Rina Gentile, Lynn Hallam, Diane Johansen, and Sharon Perkins for their assistance in the typing of the numerous drafts of this report. And finally, we are grateful for the help provided by Carol Franco and others on the Ballinger staff. Funds for the study were contributed by the Bydale Foundation, the Catherine Davis Trust, the Ottinger Foundation, the Rockefeller Brothers Fund, the Stern Foundation, and the nearly 90,000 national UCS sponsors who continue to give us the organizational capacity to carry out work of this kind.

Energy Strategies:
Toward a Solar Future

Chapter 1

Introduction and Overview

Our civilization is founded on coal, more completely than one realizes until one stops to think about it. The machines that keep us alive, and the machines that make the machines, are all directly or indirectly dependent on coal. In the metabolism of the Western world the coal miner is second in importance only to the man who ploughs the soil. He is a sort of grimy caryatid upon whose shoulders nearly everything that is *not* grimy is supported.

—George Orwell, *The Road to Wigan Pier* (1937)

INTRODUCTION

Energy is the ability to do work, to bring about change and movement, to activate. It is involved in all of the physical processes of the natural world, from the genesis of the galaxies to the growth of individual cells. Modern civilization is founded today, to update Orwell's statement, primarily on oil and natural gas in addition to coal—conventional fuels that surely will be replaced by other energy forms in the future. Nonetheless, no one would dispute his eloquent characterization of the role of energy in the development and sustenance of advanced industrial societies. It is largely a result of the introduction of technologies capable of extracting and harnessing immense quantities of energy that modern societies now accomplish tasks of tremendous scope and complexity, greatly exceeding the capacity of unaided human and animal labor, that provide hundreds of millions of people with unprecedented material prosperity. Energy is indispensable for industrial, agricultural, commercial, and personal uses of all kinds. It provides heating, cooling, lighting, and transportation. It is needed for acquiring the vast array of minerals, raw materials, foods, and manufactured products upon which modern societies depend. Indeed, the availability of large reservoirs of readily extractible fuels has allowed the United States to develop the largest gross national product of any nation in the world, with unparalleled supply of material goods, the greatest personal mobility, and the capacity—not yet fully realized—for providing adequate food, clothing, shelter, and employment for each citizen.

After years of seemingly endless supply, the United States now finds itself threatened with the unprecedented prospect of chronic energy shortages. The nation depends on irreplaceable, nonrenewable fuels for about 95 percent of its total energy supply. The exploitation of these resources has made possible great leaps in energy and economic growth. However, the rapid depletion of our highest quality energy resources has extracted a toll that we are only slowly coming to appreciate. For decades now, the nation has been living off its energy "capital" rather than off its energy "income." Our principal energy sources, oil and natural gas, are both becoming increasingly difficult, as well as costly, to produce. Each barrel of oil or cubic foot of gas consumed irreversibly diminishes the national supply and simultaneously adds to the cost of finding and producing new oil and gas. It is thus apparent that current patterns of energy consumption cannot be sustained indefinitely and, in fact, cannot be continued much longer, if at all. A further complication stems from the fact that nearly half the oil consumed in the country, or almost one-fourth of the total U.S. energy supply, is imported from abroad. We thus live with the inevitable prospect of ever-increasing prices for this imported oil plus the constant specter of abrupt cutoffs of supply.

Following shortages of crude oil, gasoline, and natural gas supplies that have developed at different times in the 1970s, a flurry of proposals to "produce our way out of the energy crisis" has consistently been revived. The UCS Energy Study has reviewed many of these energy expansionist programs and has found them to be generally reckless, as well as ultimately unworkable. The intended purpose of this study is to provide a more balanced appraisal of future U.S. energy needs and to suggest the most suitable means of going about meeting them.

It has long been apparent to us that in order to find our way out of the current morass of difficulties, it is first necessary to identify, at least in a rough way, a desirable long-term goal. Therefore, our analysis has been guided by a long-range perspective that also has a powerful bearing on the present situation. A principle conclusion is that the United States could make much better use of the energy it now consumes simply by taking advantage of currently available and economical measures to improve energy productivity. Indeed, the energy saved through technical efficiency improvements can, over the next few decades, make a contribution to the nation's supply far exceeding that possible through expansions of energy supply. Energy efficiency is not a short-term fix, however. Indeed, energy savings achieved through efficiency improvements will continue to accrue over time. Moreover, we find that the United States is, surprisingly, not "running out of energy." While certain fuels are inexorably destined to become progressively scarce, there are bountiful supplies of energy, as yet largely untapped, from sunlight, wind, plants, rivers, and oceans that are available on a sustained and renewable basis. Compared with cheap fossil fuels, these sources are difficult and expensive to tap. Yet these are sources to which the nation must ultimately turn in order to prevent the periodic reappearance of energy crises at intervals of increasing frequency. Before elaborating further on our proposed long-range solution to the nation's energy dilemma, we will back up somewhat to briefly explain how the current set of difficulties came into being.

ROOTS OF THE CURRENT DILEMMA

The United States has maintained impressive energy growth throughout the past century (see Figure 1-1). Since 1850, consumption of energy has been rising at a rate of nearly 3 percent per year, while the population has grown at a rate of less than 2 percent per year. Over this period, per capita energy use has roughly trebled, while total energy use has increased more than thirtyfold. One of the peculiar features of exponential growth is that a quantity growing at a constant exponential rate will continue to double over fixed intervals of time. From 1960 to 1973, growth in energy use accelerated to an average of 4 percent per year—a rate at which use doubles every seventeen years. More interesting is the fact that total energy use in the United States doubled from 1950 to 1972, increasing as much in that twenty-two year period as in the preceding one hundred seventy-five year history of the nation. The use of electricity, along with the industry that produces it, has grown far more rapidly than total energy use: 7 percent annually for over seventy years, doubling nearly once a decade (see Figure 1-2). In this period, electrical consumption has increased by more than one hundredfold. Growth in the 1950s and 1960s was nearly 8 percent

Figure 1-1.   Energy Consumption in the United States (1850-1978).

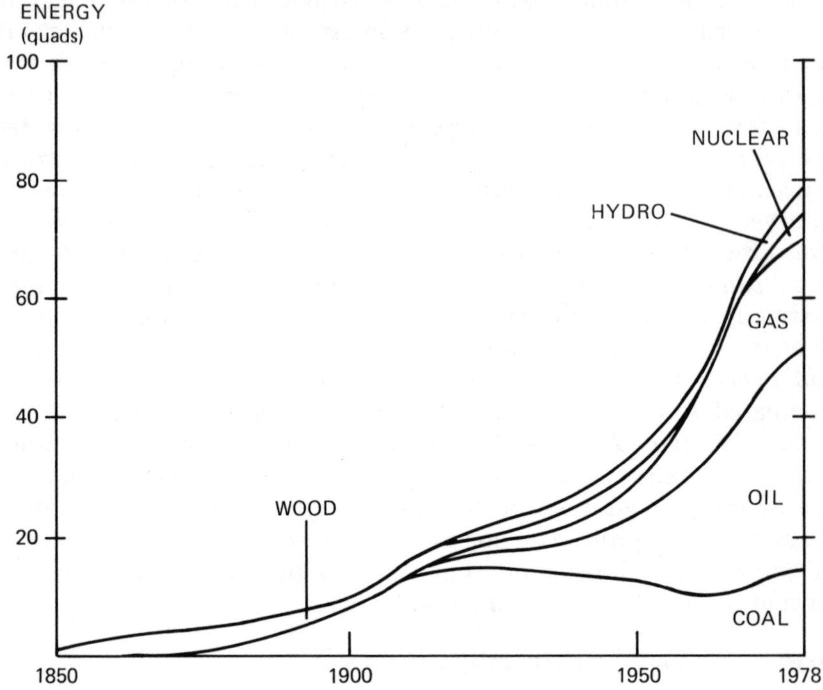

ENERGY
(quads)

Source: (DOI 1975: DOE 1979a)

per year, corresponding to a doubling of consumption every nine years. The tremendous growth in electrification can be seen from the fact that the amount of energy now used each year in the United States merely to power our air conditioners is greater than all the energy used by the nation in 1850.

The end result of these patterns of increasing energy consumption is a present level of U.S. energy use that is, by any standard, enormous. The United States, with only about 6 percent of the world's population, accounts for roughly 30 percent of worldwide energy use. The average U.S. citizen consumes approximately one hundred times more energy in a year than the average Nigerian (Erhlich et al. 1977). U.S. energy consumption now stands at a level of nearly 80 quads. A quad, the popularly unfamiliar unit of energy used by specialists, stands for 1,000,000,000,000,000, British thermal units. (A single Btu is the amount of energy required to raise the temperature of one pound of water by one degree Fahrenheit.) Perhaps more graphically, we could say that the current U.S. level of energy

Figure 1-2. Electricity Consumption in the United States, (1900-1978).

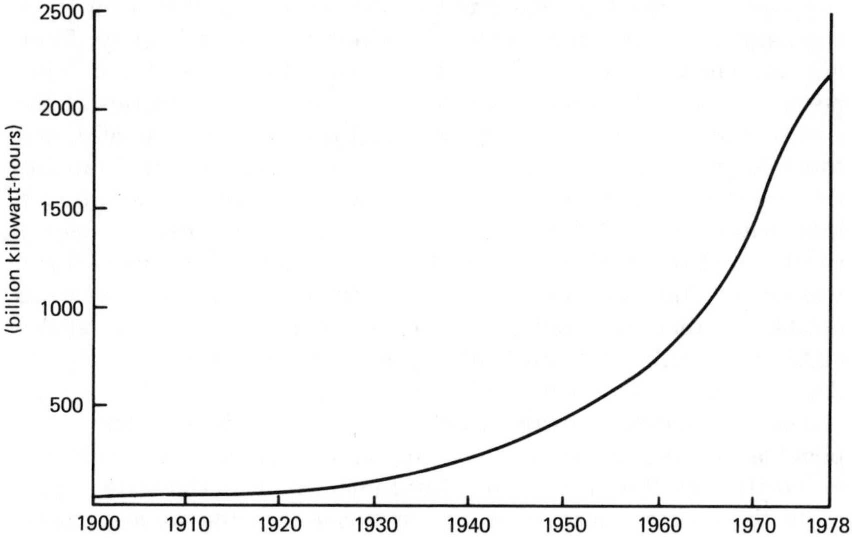

Sources: (Darmstadter 1971; DOE 1979a)

use represents the annual equivalent of 14 tons of coal or 2,600 gallons of oil for every man, woman, and child in the nation. This is indeed a prodigious supply of energy.

To meet its energy needs over the past century, the United States has relied on a changing mix of energy forms. Coal became the dominant energy source for the United States in the late 1800s, supplanting the earlier energies of wood, water, and wind. Coal production peaked around 1920 and has since fallen off slightly. Over the past fifty years, industrial societies have come instead to depend on oil and natural gas. The United States presently depends on these two fuels for about three-quarters of its energy supply, up from one-sixth in 1920. Much of this growth has taken place since 1940, with oil consumption increasing fivefold and gas eightfold since then. The dependence on foreign oil imports has also grown steadily during the same period: the United States, which was a net exporter of oil until 1948, currently imports close to 50 percent of the oil it consumes. The combined use of oil, gas, and coal presently accounts for more than 90 percent of national energy use, with the remaining requirements supplied by hydroelectric power, nuclear power, and wood.

Coal, oil, and natural gas are fossil fuels. They were formed by complex geological processes from the remains of the plant life of hundreds of millions of years ago. For all practical purposes, the supplies of these fuels are fixed in the sense that they cannot be regenerated on any time scale of interest to human society. They are nonrenewable resources: burning these fuels depletes a finite resource that will never be re-created or restored. Estimates of the size of the world's remaining fossil fuel resources vary greatly, and there is great uncertainty in the estimating procedures. Nonetheless, there is little doubt that current rates of consumption are such that within a period perhaps as short as several decades severe shortages of some of these fossil fuels, especially oil and natural gas, will occur. Thus resources that took hundreds of millions of years to create and that were only harnessed for society's use on an appreciable scale within the last century, will for the most part be rapidly depleted within the space of a few generations.

The tremendous growth in energy and electricity use was made possible by the low and stable energy and electricity prices that prevailed through the early 1970s. Government policies did little to promote the efficient utilization of our highest quality oil and gas resources and in fact contributed to quite the opposite effect. Among the policies that have facilitated rapid energy growth were direct subsidies to energy-producing industries, pricing regulations that have held fuel and electricity prices below true replacement costs, depletion allowances, foreign tax credits, and other tax programs highly favorable to energy and utility companies. In addition, consumption has also been encouraged by "promotional" rate structures employed by gas and electric utility companies that have provided a discounted rate to large users. Another major factor in the traditionally low prices of energy and electricity stems from the longstanding failure to account for the significant environmental and health costs of energy production and use by failing to require energy producers and consumers to adhere to satisfactory environmental protection standards.

The impending exhaustion of the world's oil and natural gas had not been widely recognized until the Arab oil embargo of October 1973. That dramatic event and the attendant quadrupling of world oil prices brusequely terminated the period of energy euphoria in the United States, Western Europe, and Japan. These countries, which had experienced an "economic miracle" of prodigious industrial expansion since World War II, gained a sudden appreciation of their prosperity's apparently precarious dependence on waning supplies of oil and natural gas. Modern economic growth has been so closely linked to relatively inexpensive and abundant supplies of fossil fuels that the maintenance of economic growth and the avoidance of economic decline appeared greatly jeopardized by the sudden appearance of an "energy crisis." Anxieties about fuel scarcity and its dire consequences were further enhanced by geopolitical complexities—the possibilities for arbitrary manipulation of the cost and availability of remaining oil supplies by a handful of oil-producing countries and for the radical transformation of international economic relationships by the huge cash transfers to those countries. Most industrialized countries quickly recognized the urgent need to develop national energy plans and policies that would safeguard their threatened prosperity and perhaps their national security.

However, in the years following the embargo, the United States failed to control oil imports and to initiate comprehensive energy conservation programs. Indeed, a general complacency with regard to energy supplies set in once again. A second jolt was received late in 1978, when the curtailment of Iranian oil production and exports, coupled with a complex chain of events already underway in this country, led to acute domestic shortages of gasoline in the summer of 1979.

The causes of the 1979 gasoline shortages were complicated, but appeared to extend more from the mismanagement of available supplies than from a fundamental scarcity of crude oil (Parisi 1979a; Bartlett and Steele 1979). Nonetheless, both the oil embargo of 1973–1974 and the more recent supply dislocations of 1978–1979 clearly reflect an inherent fragility underlying the U.S. energy system.

The crux of the problem stems from the fact that United States depends on depletable, nonreplenishable fuels for the vast majority of its overall energy supply. Unfortunately, necessary quantities of these fuels will not be domestically available for long. Oil and natural gas, in dwindling supply in the United States, are becoming progressively more difficult and costly to produce. Oil shortages are particularly threatening because liquid petroleum products singlehandedly fuel virtually the entire U.S. transportation sector. Domestic production of oil in 1978 was below that of 1973, despite accelerated drilling during the intervening years. U.S. oil reserves reached their peak value in 1970 and have been falling ever since (Parisi 1979b). Neglecting recent finds on Alaska's North Slope (which cannot be recovered until a gas pipeline is built), natural gas reserves have similarly declined since the peak year of 1967. In contrast to oil, there is temporarily a "glut" of gas on the U.S. market. Whether this surplus will last more than a few years, however, remains an open question. In any event, it seems clear that domestic producers will be hard pressed to maintain production at current levels over the next decade or two. Although a lot of new gas may ultimately be discovered under the Gulf of Mexico and other deep deposits, no one yet knows whether recovering it will be technically, economically, or environmentally feasible. At a minimum, it is safe to say that deep-lying gas will cost considerably more to produce than gas from already tapped deposits, additional evidence that the era of cheap energy is rapidly and irreversibly drawing to a close.

The other critical component of the nation's energy-related difficulties, which is the major contributor to recent energy price increases and perhaps the overriding concern for the short run, is the heavy U.S. dependence on imported oil. The tens of billions of dollars we spend each year on foreign oil are a significant factor in the presently imposing balance of trade deficit. A massive dependence on foreign oil has other adverse economic consequences. Large, sudden increases in the price of oil can destabilize the economy, aggravate inflation, dampen economic growth, and even help to induce a recession. The price of oil obtained from the Persian Gulf has in fact increased more than tenfold in the past decade. Within just the first six months of 1979, the price of OPEC oil rose by about 50 percent. There are political repercussions as well. Our national security now rests on an assured supply of oil. Until U.S. dependence on imported oil is reduced, the nation will remain vulnerable to continued price rises and to the possibility of supply interruptions owing to political instabilities in, or hostile intent by, producing nations. Furthermore, the potential for international conflicts will heighten as uncontrolled appetites for oil intensify competition between nations for increasingly scarce world oil supplies.

THE CONVENTIONAL RESPONSE TO THE ENERGY "CRISIS"

The official energy plans hastily formulated in the United States in the years immediately following the Arab oil embargo shared common premises and contained a generally uniform set of proposals. With names such as "Project Independence," the new plans all involved heroic efforts to acquire more energy. A major premise underlying these proposals was the widely held belief that U.S. energy consumption, and hence energy supplies, must continue to expand in order to avoid economic stagnation. These proposals called for extended oil and gas drilling, including further offshore exploration and development of Alaska's North Slope; the opening of vast areas in the western and Plains states to the strip mining of coal; the development of costly processes for converting coal and western shales to liquid and gaseous fuels; and the construction of very large numbers of nuclear-powered electric generating stations as part of a program to increase the electrification of our energy supply system.

A strikingly similar set of proposals emerged once again in 1979 in response to the sudden reappearance of the energy "crisis." The plans unveiled by President Carter on July 15 and 16, 1979, again placed near-total emphasis on expanding energy supplies rather than on promoting improvements in the efficiency of energy use. The increase in energy supplies was expected to come from coal, oil shale, and nuclear power. At the heart of the plan was the proposed establishment of a major industry for the production of synthetic fuels, predominately from coal and shale resources. Indeed, an ambitious goal of producing the equivalent of 2.5 million barrels of synthetic oil per day (corresponding to about 14 percent of current U.S. oil consumption or slightly under 7 percent of the total U.S. energy supply) was set for 1990. The costs of implementing the president's program would be staggering—a minimum of $142 billion to be spent over a ten year period, financed almost exclusively through a proposed "windfall" profits tax to be imposed on oil companies. The way in which this money is to be allocated leaves little doubt as to the administration's priorities: $88 billion (or more than 60 percent of the total) would go toward the development of synthetic fuels, while only $18.5 billion (about 13 percent of the total) would be used to upgrade mass transit systems, improve automobile fuel efficiency, and promote conservation in oil-heated buildings (*New York Times* 1979).

In addition to the formidable price-tag, large-scale energy acquisition ventures along the lines proposed would pose massive threats to the environment and human health. Major increases in coal production would inflict severe damage to the land. Large tracts of western land would have to be strip mined to achieve contemplated production goals. The Department of Energy, for example, is predicting that coal production will double by 1990 (DOE 1979b). At present there are no satisfactory techniques for reclaiming strip mines in the arid western regions where the most abundant supplies of coal lie. The cumulative production of coal from underground mines in the East would also cause substantial land damage from subsidence—the surface collapse of land that has been previously undermined. Large amounts of pollutants would be dumped into the air by coal-fired power plants. Demands for cooling water, a scarce commodity in many regions, would be awesome if cooling ponds or wet cooling towers were employed in these plants.

Severe impacts would also result from the conversion of coal to synthetic fuels. To produce significant quantities of synthetic oil and gas, coal mining, with its attendant impacts, would have to be greatly increased. Owing to fuel conversion efficiencies no better than 60 to 70 percent, roughly 1.5 tons of coal would have to be mined to produce synthetic oil with the energy content of 1 ton of coal (Wilson 1977). In addition to the disruptive effects of mining, both coal gasification and liquefaction plants would require large land areas, emit considerable quantities of pollutants to the air, generate large volumes of solid wastes, and consume prodigious amounts of water.

A major commitment to oil shale development would bring similar consequences. Assuming that oil shale were produced by the established method of mining and surface "retorting" rather than yet to be developed, underground "in situ" techniques, vast amounts of earth would have to be moved and processed—approximately 7 tons to deliver as much energy as is contained in 1 ton of coal. A one million barrel per day shale industry would involve moving a billion tons of rock and residue each year—roughly comparable to the total amount of material excavated in the building of the Panama Canal (Gillette 1979). Prodigious volumes of wastes—about a half billion tons per year for a one million barrel per day industry (Schuyten 1979)—would be generated. "Spent" shale cannot simply be returned to the earth, because in the process of crushing, its volume is increased by about 20 percent over its original size. Present plans call for its disposal in western canyons. Water consumption would also be substantial. Roughly two to three barrels of water would be consumed, and consequently polluted, for every barrel of shale oil produced; the production of one million barrels of oil per day would annually require thirty to forty-five billion gallons of water per year—equivalent to almost 10 percent of the Colorado River water now consumed in the western United States (Gillette 1979).

Longer term climatic considerations also militate against a greatly expanded use of coal and synthetic fuels. Scientists are now in general agreement that cumulative additions of carbon dioxide to the earth's atmosphere from a continually expanding consumption of fossil fuels will lead to a significant warming of the world's climates. A 1979 report to the Council on Environmental Quality indicates that the resultant global heating could be pronounced before the end of this century (Woodwell et al. 1979). Although the precise effects of the climatic alteration are difficult to predict, there is little reason to believe that they will be beneficial to mankind. Climatic zones will shift, threatening the stability of food supplies. With sufficient warming, the polar ice caps would melt, raising sea levels and inundating low-lying coastal cities. While the burning of all fossil fuels produces carbon dioxide, some are worse than others in this respect. Unfortunately, coal and coal-derived synthetic fuels are among the worst offenders, releasing roughly twice as much carbon dioxide per unit of energy produced than an equivalent amount of natural gas (Woodwell et al. 1979). Thus, there is good reason to question the goal of increased coal and synthetic fuel production as a means of offsetting future shortfalls in oil supplies.

Nuclear energy, another major component of official scenarios for the future U.S. energy supply network, has its own troubling drawbacks. The risks of catastrophic accidents have been extensively studied by scientists and, though real, are difficult to quantify in an entirely satisfactory way. While there is no disagreement about the fact that nuclear power plants cannot explode like nuclear weapons, informed technical opinion is divided on whether current nuclear power plant safety defenses are adequate to prevent other types of damaging accidents. These accidents, which could disperse massive amounts of radioactivity from a nuclear power plant into the surrounding area, have not arisen to date in the country's limited commercial nuclear power experience, but the long-term chances of operating nuclear plants while avoiding such potential calamities remain an open question. A major catastrophe was fortunately averted during the March-April 1979 accident at the Three Mile Island nuclear plant. The event, however, did little to engender confidence in the adequacy of reactor safety or in the competence of federal regulators and plant personnel. Also unresolved is the complex problem of radioactive waste disposal, a grim legacy left from the nuclear program for future generations to contend with. No satisfactory radioactive waste disposal technology has yet been developed despite the nearly three decades of work in this area by the federal government. Nor is there a convincing means of safeguarding the by-product plutonium from nuclear plants; this material, a prime ingredient for nuclear weapons manufacture, could create grave national security problems if inadequately protected from theft or diversion.

It is clear that a strategy emphasizing increased production of coal, synthetic fuels, and nuclear energy will do little to remedy the fundamental problems stemming from a reliance on exhaustible, irreplaceable, and hence, inflationary energy supplies. At best, such a strategy could merely postpone—and thus make far more difficult—the deadline for a transition to truly renewable energy supplies. In the short run, in light of the considerable costs and risks involved, a massive expansion of conventional means of energy production will almost certainly exacerbate, rather than alleviate, our present energy-related difficulties. In fact, such a program can accurately be characterized, as David Brower has observed, as one of "strength through exhaustion."

In spite of the unquestionably adverse consequences of implementing these plans, the conventional wisdom maintains that current energy policies are justified by the overriding "need for more energy." The propensity to expand conventional fossil fuel and nuclear power production is based on the further belief that alternate energy resources cannot be readily developed and exploited. While alternative, renewable sources of energy may some day make a substantial contribution to the national energy supply, that day, according to this view, is presumed to be far into the future. Consequently, a concerted effort to foster the early introduction of even the most promising new power sources is not a major component of the current U.S. energy strategy, and a correspondingly low priority has been attached to research and development efforts in this area.

The "need for more energy" and the status and potential role of alternate energy sources were the principal issues addressed by the UCS Energy Study. We have found the "conventional wisdom" to be wrong in dealing with both issues. On the contrary, we find the need for ever-increasing amounts of energy can be obviated by technically and economically feasible improvements in the efficiency of energy use. Available improvements in energy efficiency can in fact have quite impressive consequences. Furthermore, we conclude that "alternative," renewable resources offer the most promising long-range energy option for the United States. Moreover, these sources can begin to make substantial contributions in this century. Our conclusions on these key issues are summarized in this overview and detailed in later chapters.

ASSESSING THE "NEED FOR MORE ENERGY"

It is generally assumed by U.S. energy planners that the already high level of energy consumption in the U.S. will, in conformance with historical trends, increase substantially in coming years. A further belief is that without constantly increasing supplies of energy, the economic well-being of the nation will be jeopardized. Official "forecasts" and "projections" of future energy needs have traditionally been based on extrapolations of historical trends. Overall energy use had steadily increased for decades and, it has been assumed, would continue to do so in the future. Electric power consumption was similarly forecast to continue its spectacular growth in coming decades. Since population growth has been slower, the projected requirements for electrical energy implied remarkable increases in per capita electric power usage. For example, in the early years of the 1970s, it was widely assumed for planning purposes that per capita electric power requirements for the year 2000 would be eight times 1970 levels—a result arrived at on the basis of extrapolations of past trends. While acknowledging that some short-term and limited reduction in energy requirements could be achieved through "energy conservation," national energy planners have taken it as a basic and irreproachable truth that continually expanding energy supplies were an absolute requirement for national economic growth and well-being.

The presumption of a strong, essentially inflexible link between constantly expanding energy supplies and continued economic growth requires careful examination. Common sense and practical observation tells us that we need a good deal of energy to run advanced industrial economies. At issue, however, is the level of required future energy supplies and the question of whether we need steadily increasing consumption of energy above the already high current levels to assure continued prosperity and economical growth. Unfortunately, in the mild hysteria that followed the Arab oil embargo, and again after the more recent "crises," there was no critical review of this cardinal issue. Government agencies merely drafted the various national energy plans on the common but unjustified assumption that the level of energy use inevitably had to rise in order to increase the national "standard of living." In truth, however, this centrally important and widely accepted dogma embodies a grossly unbalanced and incorrect view of our economy's use of, and dependence upon, energy resources.

Energy, as we have noted, is obviously a vital input to our industrial economy. Energy, however, should not be regarded as some kind of supercommodity around which economic life revolves: it is but one of a broad array of useful and vital commodities. Potable water, fertile land, and basic raw materials, not to mention supplies of labor and equipment, are as necessary as energy sources in the creation and maintenance of the national health and well-being. Energy is no more essential than any of the other essentials upon which we depend, and it is fallacious to place energy sources in some transcendental category that implies that they escape the customary laws of economics and must be supplied in ever-increasing amounts.

The level of energy use, far from being determined outside normal economic considerations, has manifestly followed the normal course. It is precisely because energy prices were so low relative to other costs—in part the result of government subsidies and pricing policies— that the present wasteful patterns of energy consumption were established. Energy prices have been so low, typically no more than a few percent of total production costs in industry or in the overall cost of living, that energy-intensive processes have been widely adopted, and little care has been taken to economize on energy use. When energy is cheap, it doesn't cost much to waste it, and it doesn't pay to spend much money to save it. As a result of consistently low prices, energy users have been led to believe that energy resources could forever expand to meet whatever demands might arise. This belief is seriously mistaken.

In fact, the productivity of energy—the amount of energy required to produce a given good or provide a given service—is flexible, rather than rigidly fixed. The substitution of labor, materials, or capital in place of energy can all modify to some extent the amount of energy needed to perform a given task. Therefore, following normal readjustment processes, increased energy prices relative to other prices will inevitably alter future patterns of energy use. To think otherwise, and to postulate increasing energy use regardless of price, is to pretend that the economy's governing decisionmaking mechanisms have become unhinged and inoperative.

The conventional forecast of an ever-swelling demand for energy, in the face of greatly increasing costs, is therefore a dubious speculation that contradicts very elementary observations about our economic processes. A more thorough and realistic assessment of future energy requirements, which systematically takes into account the type of energy saving efforts mandated by the currently higher price of energy, offers a remarkably different prognosis. So many obvious and economically advantageous techniques are now available for increasing energy efficiency and productivity—that is, for raising the level of economic activity supported by a unit of energy—that the United States should be able for decades to come to maintain and indeed to advance its level of material prosperity without substantially higher—and very possibly lower—levels of energy use.

Astonishingly little effort has been devoted to the efficient utilization of the vast quantities of energy consumed in the United States. Electric power plants, for example, typically convert only about a third of the heat generated in their boilers or nuclear reactors into electricity; the leftover energy is normally dumped into the air or into a neighboring body of water as "waste" heat. Homes are generally so poorly insulated and constructed that a large fraction of the heat put into them quickly leaks out. Industrial complexes typically burn fuels to produce process steam and "import" electricity from the local utility company, when in fact, substantial fuel savings could be achieved by combining the two processes within the industrial plant. Today's automobiles, with their overly powerful engines, typically consume at least twice as much energy as would be used by lighter, more efficient vehicles.

The cumulative reduction in energy requirements stemming from systematic efforts to improve end use energy efficiency has been established in a number of recent, independent studies. These include the Energy Policy Project of the Ford Foundation (1974), American Physical Society (1975), Ross and Williams (1976), Widmer and Gyftopolous (1977) and Stobaugh and Yergin (1979). These investigators have focused on known and established methods, and their findings do not presuppose advanced energy-saving techniques yet to be developed—a fact that makes their conclusions exceedingly conservative when applied to long-range projections of energy needs. The general finding of these studies is that there are straightforward technical means that could roughly cut U.S. gross energy needs in half. Far from being an unrealistic target, efficiency improvements on this order would merely reduce average per capita energy requirements to levels already characteristic of the more efficient countries of Western Europe, such as Sweden, Switzerland, and West Germany, whose citizens enjoy living standards comparable to those in the United States. Recent experience gained in U.S. industries is even more convincing. For example, from 1973 to 1978, net industrial output increased by 12 percent while total energy use actually declined by 6 percent (White House 1979). This powerfully contradicts the common assertion that constantly increasing supplies of energy are an absolute requirement for sustaining U.S. economic growth. If energy were used with steadily increasing efficiency in years to come, our current levels of consumption would continue to support economic growth for decades.

Chapter 2 of this study offers a detailed elaboration of the more realistic appraisal of U.S. energy needs that we have summarized here. In it we examine the current structure of energy use in each sector of the U.S. economy and assess the degree to which the efficiency of energy use could be improved by purely technical measures. Considered, for example, are the energy savings that can be achieved by improving the design of buildings, machines, appliances, and transportation vehicles; by employing more efficient industrial practices; and by using the waste heat normally rejected in the generation of electricity. In addition to technical efficiency questions, we investigate how population growth and increases in the average standard of living affect future energy needs. This exercise involves what is called a "parametric analysis"—the attempt to estimate the impact of variable factors (such as population growth, standard of living, and efficiency) on the overall demand for energy. The conclusion that emerges from this analysis is clear: the United States can provide a high level of economic well-being for all its citizens without the high levels of gross energy use typically assumed in official projections of future energy requirements.

In contrast to the "conservation" programs typically endorsed by the federal government, it is noteworthy that our analysis of future energy needs incorporates technical efficiency improvements rather than austerity—for example, lowered thermostat settings, doing without appliances, or reduced travel—to minimize the growth in energy consumption. No radical transformation of society or reversion to the lifestyles of our grandparents is required to hold down growth in energy use. Technical sophistication, rather than sacrifice and curtailment, provides the key to our energy efficiency strategy.

Energy efficiency is indeed a major energy resource. For example, a mere 3 percent reduction in the amount of energy needed to power the U.S. economy would save as much energy as is contained in the oil now flowing in the Alaskan pipeline—about 1.2 million barrels per day. If the country wished to avoid a dependence on OPEC oil supplies, this could be accomplished by a 15 percent increase in national energy efficiency, which would save as much oil as we currently import from OPEC nations. A 25 percent improvement in overall efficiency would save more oil than we currently import from all foreign suppliers. (It may, however, be to our advantage in some cases to consume foreign oil and save our own.) A program of industrial co-generation of process steam and electricity could economically produce at least twice as much electricity as is currently provided by all of the nation's nuclear power plants. Important savings could be realized because industrial co-generation requires only half as much energy, in addition to that required for industrial process steam and heat applications, as is normally used to generate electricity.

In terms of extending our remaining oil and natural gas resources, a systematic national program to improve energy efficiency would have a dramatic impact. If the efficiency of energy use were gradually raised nationwide so that we employed 25 percent less energy in the year 2000 and 50 percent less energy in the year 2050 to support our current level of well-being, cumulative energy savings through 2050 would amount to roughly 3000 quads, exceeding the combined total of our estimated oil and gas resources. Recent new international oil and gas discoveries, such as have been announced by Mexico, might also eventually help extend the future availability of oil for the United States.

We can conclude, therefore, taking both a broad view of energy and the economy and a particularly hard look at specific energy-saving techniques, that the energy supply needs of the U.S. have been grossly exaggerated by the nation's energy planners. It follows from this that the justification for some of the risky energy acquisition ventures now underway is highly questionable and that the useful life of remaining oil and gas resources can be substantially "stretched out," leaving sufficient time to plan for and to begin to implement the introduction of new, benign energy technologies. Instead of working furiously to expand conventional energy supplies, with all the punishing environmental and public health repercussions that this implies, the United States can set out on a more systematic and rational basis to build a truly stable long-term energy supply system.

EVALUATING LONG-TERM ENERGY SUPPLY OPTIONS

Irrespective of what improvements in energy efficiency can be achieved, it is clear that our present sources of energy will ultimately have to be replaced by others. Within about seventy-five years or less, the United States will no longer be able to maintain its dependence on fossil fuels. Domestic supplies of oil and gas will be largely exhausted by that time, and the remaining amounts globally will be very expensive to extract. Coal will still be relatively abundant, but its use will almost certainly have to be restricted by environmental and climatic considerations, as noted previously. It is therefore urgent that the nation largely complete the transition to a renewable energy system by the middle of the next century, if not sooner.

For the long term, the United States will have to choose among only three sources of energy—solar energy (including indirect forms such as wind, biomass, and hydropower), breeder reactors, and nuclear fusion. Our major finding is that solar energy utilization offers uncontestable environmental and social advantages over the others and that a workable strategy exists to make solar energy the basic supplier of the nation's energy. Breeder reactors, with their formidable inventories of plutonium, would pose unparalleled risks to society in a large-scale program and would offer little respite from the array of problems besetting the current generation of nuclear reactors. Nor can nuclear fusion now be depended upon to rescue us from the energy "crisis," as its technological feasibility still remains to be demonstrated. In light of the difficulties confronting the other major, long-term energy options, it is worth taking a hard look at the potential for meeting all future energy needs from the sun.

A SOLAR ENERGY FUTURE

While solar power is now widely advocated as a future energy source, no overall "systems" approach has been offered that takes account of the subtle, practical problems of utilizing solar technologies to produce energy in all the forms required to meet the diverse needs of an advanced industrial economy. Nor has there been a systematic, critical evaluation of various solar options to distinguish workable applications from more marginal proposals that may either fail technically or produce unacceptable side effects.

In the UCS study, we have assessed the major solar energy options for the United States. Included is a review of the potentially available renewable energy technologies: solar thermal, photovoltaic, wind, biomass, hydroelectric, satellite power, ocean thermal energy conversion, tidal power, wave power, ocean current, and salinity gradient energy systems. A guiding factor in this analysis has been a "systems" approach: this recognizes that it is not enough merely to be able to capture solar energy in one of its many forms; to be of any great usefulness, energy must also be made available in the proper form, at the right time and place, and at an affordable cost. Therefore, all components of an overall energy supply system—including energy collection, conversion, distribution, and storage facilities—have been carefully examined.

The sun is the world's most abundant, inexhaustible source of energy. It drives the global climatic systems, powering the evaporation-precipitation cycle of the biosphere; supplying the kinetic energy of the winds, waves, and ocean currents; and causing the temperature gradients within the oceans. Solar energy sustains life on the planet by supporting photosynthetic plant growth and thereby providing for all the food we eat. Firewood—the major source of energy in the United States through the nineteenth century and still the principal energy source for much of the developing world—is an indirect form of solar energy. Even the energy contained in fossil fuels—which provide over 90 percent of the current U.S. energy supply—was initially derived from the sun. Thus, it is incorrect to regard the sun as a "new" and "exotic" energy source. On the contrary, recent proposals to harness solar energy merely imply extending the range of applications for what already constitutes our most significant energy resource.

The virtues of the solar resource are readily apparent: it is renewable, nonpolluting, abundant, and widely distributed. On an annual basis, approximately 44,000 quads of sunlight fall on the U.S. land mass; in comparison, total energy use in the United States is less than 80 quads per year. Sunlight is a high quality energy source that can be readily converted to various grades of energy, depending on the application at hand. Solar energy conversion technologies are inherently flexible and can be constructed on a variety of scales ranging from small-scale residential units to systems designed for large-scale metropolitan and industrial uses. The distributed nature of the solar resource makes possible the collection of energy at or near the point of end use, in some cases obviating the need for energy transmission with its attendant costs and energy losses.

The most important questions concerning solar energy utilization thus relate not to the magnitude of the solar resource but to whether energy can practically and economically be made available in appropriate forms at the required times and locations. There are several obstacles to harnessing the sun's energy that must be considered.

Solar energy is both diffuse and intermittent, posing practical difficulties for its utilization. The diffuse nature of solar radiation means that considerable amounts of land must be devoted to energy collection. Nonetheless, land use should not pose an insuperable barrier to the widespread utilization of solar energy. With solar collectors operating at thermal efficiencies of 40 percent, less than 0.5 percent of the total U.S. land area would need to be devoted to energy collection in order to provide for the current level of energy demand. This is less than half the amount of land presently covered with roads. Devotion of land to solar energy acquisition would not necessarily preclude other uses. For example, solar energy systems could, in many cases, be integrated into buildings and other structures. Similarly, windmills could generally be sited on agricultural and ranch lands without interfering excessively with other operations. In any event, the land impacts would be much less punishing than those associated with coal mining, for example.

The intermittent nature of solar energy means that energy storage will ultimately be essential in a solar-based energy economy to ensure the availability of energy upon demand. As discussed extensively in Chapter 5, the need for storage should not pose a significant barrier to the near-term utilization of solar energy. Nonetheless, the development of suitable storage technologies and the integration of solar technologies with conventional energy supply systems do pose one of the major technical challenges remaining for a long-term solar economy.

Many of the solar technologies that will ultimately be needed are still at early stages of development, and the economics are generally unfavorable at present. In the vast majority of cases, however, no major technical breakthroughs are needed for commercialization. The current high cost of most solar devices constitutes the most significant barrier to their utilization. The general economic competitiveness of solar technologies can be dramatically improved by redirecting federal policies in such a way as to offset the vast subsidies currently afforded to conventional energy sources. In many cases, all that is needed to bring the costs of solar devices down is to remove barriers to use and thereby create a market sufficient to support the introduction of mass production techniques.

Although the solar resource is prodigious, there are practical limits to the amount of energy that can be extracted from it without adverse consequences. While renewable energy sources can often be harnessed on a small scale with negligible environmental impact, major climatic shifts could ensue if exploitation were to proceed beyond a certain threshold level. Furthermore, solar technologies are not equally benign. While solar energy can typically be utilized with far lower environmental impacts than with most conventional energy sources, some solar technologies are markedly inferior to others in this respect. Among the most favorable means of solar energy utilization are "passive" and "active" solar heating and cooling; photovoltaic, wind, and solar thermal-electric systems; and the conversion of organic residues and wastes to "biofuels." However, major environmental questions mar the future desirability of microwave-beaming solar power satellites, floating ocean thermal energy conversion plants, and large-scale, monocultural biomass "plantations."

On the basis of a detailed weighing of solar power options, this study has identified a specific solar strategy capable of providing energy in the requisite forms needed by our advanced industrial economy. A solar energy future implies diversity: no single solar technology can solve all our energy problems or satisfy all our needs. Thus, a broad mix of solar technologies—suited both to the natural energy income indigenous to a specific region and to the required end use energy form—will be needed. In order to obtain an optimal matching of the most promising energy supply technologies with appropriate end uses, we have carefully assessed future requirements for thermal energy, electricity, and liquid and gaseous fuels in the residential, commercial, industrial, and transportation sectors.

We conclude that with a shrewd selection and integration of solar energy supply, storage, and transmission technologies, the full range of future U.S. energy requirements can be met. We believe that a transition to a wholly solar-powered economy could be largely completed roughly by 2050. Moreover, solar energy can make an important contribution to the nation's energy supply in this century, providing anywhere from 15 to 33 percent of total primary energy use by the year 2000, depending on the commitment made both to improving energy efficiency and to deploying solar technologies. The technical basis for these seemingly optimistic conclusions is provided in Chapters 4 through 6.

It is now abundantly clear that the world has entered a period of chronic energy shortages that will continue until mankind has learned to harness energy from renewable sources. What seems certain, at least for the foreseeable future, is that energy, once cheap and plentiful but now expensive and limited, will continue to rise in cost. Historic patterns of ever-increasing consumption cannot continue. Moreover, current federal policies to expand energy supplies cannot be implemented without almost certainly ruinous effects. In the face of these mounting difficulties, a long-term strategy emphasizing major improvements in energy productivity (with an attendant reduction in energy growth) and a reliance on renewable energy derived from the sun emerges as the clearest and most sensible solution to the great challenge we and future generations face. Changes will inevitably occur as supplies of our present fuels dwindle and disappear. We believe that the strategy outlined in this study can help ease the burden of transition to a benign and sustainable energy future for the United States.

REFERENCES

American Physical Society. 1975. "The Efficient Use of Energy: A Physics Perspective. A Report of the Summer Study on Technical Aspects of Efficient Energy Utilization." New York. January.

Bartlett, D.L., and J.B. Steele. 1979. "How Blunders Cut Our Supply of Gas." *Chicago Tribune*, July 1, sec. 2, pp. 1–2.

Darmstadter, J. 1971. "Energy Consumption: Trends and Patterns." In *Energy, Economic Growth, and the Environment.* Baltimore: Johns Hopkins University Press, April.

Department of Energy (DOE). 1979a. "Annual Report to Congress." Vol. 2. Washington, D.C.: Energy Information Administration, DOE/EIA-0173/2.

——. 1979b. "Energy Supply and Demand in the Midterm." Washington, D.C.: Energy Information Administration, DOE/EIA-0102/52.

Department of the Interior (DOI). 1975. "Energy Perspectives." Washington, D.C.: Government Printing Office, February.

Energy Policy Project of the Ford Foundation. 1974. *A Time To Choose— America's Energy Future.* Cambridge, Mass.: Ballinger Publishing Company.

Erhlich, P.R.; A.H. Ehrlich; and J.P. Holdren. 1977. *Ecoscience: Population, Resources, Environment.* San Francisco: W.H. Freeman and Company.

Gillette, R. 1979. "Long Road To Synthetic Fuel." *Boston Globe,* July 16, p. 1.

New York Times. 1979. "How President Intends To Use Oil Profit Tax." July 18.

Parisi, A.J. 1979a. "Oil Data Suggest Small Shortage Became Crisis From Mishandling." *New York Times,* July 18.

———. 1979b. "Nation's Energy Chain: Fuel Sources Intertwine." *New York Times,* July 16.

Ross, M.H., and R.H. Williams. 1976. "Energy Efficiency: Our Most Underrated Energy Resource." *Bulletin of the Atomic Scientists* 32 (November): 30–38.

Schuyten, P.J. 1979. "The Synthetic Solution: The Rub is the Cost." *New York Times,* July 15.

Stobaugh, R., and D. Yergin. 1979. *Energy Future: Report of the Energy Project at the Harvard Business School.* New York: Random House.

White House. 1979. "Fact Sheet on the President's Program." Washington, D.C.: Office of the White House Press Secretary, April 5.

Widmer, F.R., and E.P. Gyftopolous. 1977. "Energy Conservation and a Healthy Economy." *Technology Review,* June, pp. 31–40.

Wilson, C. 1977. *Energy: Global Prospects 1985–2000.* New York: McGraw-Hill Book Company.

Woodwell, G.M., et al. 1979. "The Carbon Dioxide Problem: Implications for Policy in the Management of Energy and Other Resources." A Report to the Council on Environmental Quality. Washington, D.C., July.

✳ *Chapter 2*

An Assessment of Energy Needs

INTRODUCTION

It is critically important to have a clear idea as to the amount of energy we will need in the future and roughly how that energy will be utilized to provide useful services. Of concern is both the quantity and quality of energy required to meet our diverse needs. Electricity, for example, is a very high quality energy form suitable for appropriate tasks such as lighting and the operation of electric motors and appliances that can take advantage of its premium quality. High temperature heat, needed to produce steam or to drive industrial processes, represents energy of somewhat diminished quality. Only the lowest grade of energy, low temperature heat, is needed for such tasks as building and water heating.

Oddly enough, a thorough assessment of the quantity and form of energy needed to perform necessary functions in the future is conspicuously absent from most energy plans. Normally, an arbitrary rate of energy growth is assumed (for reasons rarely subjected to critical scrutiny), and attention is quickly focused on the "more interesting" question of how to go about meeting the projected demand by bringing into play a battery of energy supply options. In contrast to the traditional approach, we view an energy needs assessment to be the logical starting point for a broader analysis of the country's energy future, deserving most thorough consideration.

In this chapter, the role of energy in the economy is addressed, and reasonable estimates are made regarding future energy needs. In particular, we scrupulously evaluate the "conventional wisdom" regarding the need for ever-increasing supplies of energy to support economic growth. Our analysis finds this assumption to be indefensible; in fact, a plethora of options exist to raise the productivity of energy that allow for the maintenance of a healthy economy without major increases in energy use. Indeed, decreased consumption may even be possible in some cases.

In estimating future energy requirements, we follow a course at odds with the standard, but faulty, practice of extrapolating from past growth rates. Instead, present patterns of energy use are examined in order to establish how future energy needs might be distributed. Next, ways of reducing energy usage through the adoption of efficiency techniques are reviewed. Also considered are the influence of population growth and increases in per capita energy use (reflecting raises in the standard of living) on future energy demand. This approach to computing future energy requirements not only leads to much lower estimates than those arrived at on the basis of simple extrapolations, but also sheds some light on the manner in which energy will ultimately be utilized at the projected levels of demand—information that is not conveyed in standard growth curves.

ENERGY AND THE ECONOMY

Energy is a critical input to the U.S. economy at myriad levels. It makes possible the extraction, processing, and manufacture of materials into finished products and allows for the transportation and distribution of these products to consumers. Over the past several decades in the United States, relatively sustained economic growth has been closely accompanied by a steady increase in energy consumption. This close historical correlation has contributed to naive assumptions about the linkage between economic growth and energy consumption.

Despite energy's vital role in the economy, it does not stand as an independent entity. In order to provide useful services, energy must be utilized in conjunction with labor, materials, and technology. To some extent, these can all be substituted for energy to achieve a desired end. Furthermore, the total cost of energy accounts for only a small percentage of the gross national product (GNP). This is presently estimated to be about 5 percent of the U.S. total (Keeny et al. 1977).

Therefore, the quantity and quality of services provided and of goods manufactured are neither determined nor limited solely by the energy input. On the contrary, the productivity of energy is a variable quantity, capable of responding to changing conditions. Hence, the interdependence between economic growth and energy consumption is not nearly as strong as many in the United States believe.

That energy use is now largely decoupled from economic growth stems in part from the profound difference between the present situation and that which prevailed from 1960 to 1970. While energy and electricity prices either declined or remained at relatively constant levels during that period, at present the development of new energy supplies and the construction of new power plants have become increasingly expensive. Owing partly to stable or declining energy prices over the past several decades, relatively little attention was addressed to increasing the efficiency of energy usage. In fact, the energy used to perform a given task—such as to heat and cool buildings and to power automobiles and aircraft—generally increased over that period (Ross and Williams 1977). In this new era of inpending fuel shortages and correspondingly higher energy prices, the notion of efficient energy use has finally gained a long overdue modicum of respectability.

Several recent studies (EPP 1974; Ross and Williams 1977; Widmer and Gyftopolous 1977) have come to conclusions that directly challenge traditionally held assumptions on the interdependency of energy growth and economic well-being. These studies indicate that technical improvements in the efficiency of energy utilization could dramatically reduce the growth in energy consumption without incurring major economic upheaval or requiring sacrifices in living standards. In fact, these analyses point out that a healthy economy could be maintained and economic growth continued at near-historic rates with little or no net growth in energy use.

Recent experience has supplied supporting evidence that is incontrovertible. For example, in 1978, growth in the U.S. gross national product averaged 3.9 percent, while energy use grew less than half as fast—at a rate of 1.9 percent (*New York Times* 1979). Similarly, energy use grew only 70 percent as fast as the GNP from 1975 to 1977 (Halloran 1979). Even stronger evidence is provided by energy efficiency gains achieved in the industrial sector, where total energy use declined by 0.1 percent from 1976 to 1977, while production increased by 5.9 percent (CEQ 1978). Over the 1973–1978 period, industrial energy use dropped by 6 percent, though output was increased by 12 percent (White House 1979).

It is important to emphasize that this proposed reduction in future energy growth could be made to occur simply by using energy more efficiently, rather than by forced cutbacks in supply or major modifications in lifestyles. In contrast to the disruptive effects of the 1973–1974 oil embargo, in which abrupt fuel shortages impaired both business and employment, the planned and gradual implementation of more energy-efficient methods and technologies can occur without major dislocations. In the majority of cases, measures designed to save energy are economically attractive today. While a nationwide effort aimed at reducing inefficiency to free up energy for other uses would involve a major financial commitment, the investments required would be substantially less than would be needed to expand supplies to a corresponding extent (EPP 1974; Ross and Williams 1977; Widmer and Gyftopolous 1977).

Other benefits would accrue in addition to the savings of capital. An aggressive energy efficiency program would open up many new job and business opportunities for the manufacture, marketing, installation, and maintenance of energy conservation equipment. The number of newly created jobs would offset the number eliminated in energy supply and other highly energy-intensive industries, which employ relatively few people and rely to a large extent on inputs of energy and capital in place of labor (EPP 1974). Furthermore, an emphasis on saving energy, instead of on developing new sources, will have an overwhelmingly positive effect on the environment, alleviating energy-induced health and environmental damage. Finally, a program aimed at curbing unproductive energy growth can help provide the additional time needed to systematically evaluate future energy supply options before premature and unwise commitments are made in terms of misspent dollars, resources, and human effort.

In light of the ostensibly compelling advantages offered by increased energy efficiency, it might appear puzzling that investments directed to energy conservation are proceeding so slowly. Progress in this area has been impeded by a combination of obstacles. Conservation runs contrary to the normal way of doing business in the U.S. economy. The wide range of direct and indirect subsidies afforded to conventional energy sources has created a distorted market that tends to inhibit conservation expenditures. Furthermore, no clear constituency exists for the advocacy of energy efficiency. Given adequate time and a "corrected" market, cost-effective conservation investments would ultimately be made. In order to facilitate a rapid boost in energy productivity, however, more overt and immediate measures are needed. A more detailed discussion of the obstacles to energy conservation and the means of overcoming them is provided in Chapter 6.

ESTIMATING FUTURE ENERGY REQUIREMENTS: THE CONVENTIONAL APPROACH

In estimating energy demand, an important distinction should be drawn between the concepts of gross energy and net energy. Gross energy represents the total energy used in the performance of a task, including losses due to avoidable or inevitable inefficiencies. Net energy represents the amount of energy actually required to perform a designated task, neglecting losses. For example, in the generation of electricity, three kilowatt-hours of primary fuel energy are typically required to produce one kilowatt-hour of energy in the form of electricity. The net energy required to illuminate a 100 Watt light bulb for ten hours would thus be one kilowatt-hour, whereas gross energy requirements would be three kilowatt-hours. Unless otherwise indicated, all energy values presented through this study can be assumed to represent gross, rather than net, energy values.

The extrapolation of historical energy growth rates has provided the basis for most projections of future energy requirements formerly made by the federal government and utility industries. More recent government projections generally depend on complex econometric models that are also of questionable reliability. An inevitable consequence of relying on extrapolative models is that an ever-increasing growth of energy demand is automatically built into future projections. For example, estimates made on the basis of an assumed 4 percent growth in energy use (AEC 1974; FEA 1974; Exxon 1975) lead to projected demand levels of more than 190 quads for the year 2000, almost two and one-half times the 1978 rate of use. Even when more "reasonable" assumptions are made (i.e., energy use growing at less than 3.5 percent per year in response to increasing energy prices), very high values are still arrived at—an energy demand of 160 quads by 2000 and 350 quads by 2025 (Keeny et al. 1977). Conventional projections of future energy demand are depicted in Figure 2-1. For the sake of comparison, considerably lower estimates arrived at by more recent studies are also included in the figure.

Similar projections have been made for the future use of electricity. Many forecasts made in the early 1970s assumed that the use of electricity would continue to grow at the historic rate of 7 percent per year, doubling every ten years (AEC 1974; Dupree and West 1972). By the turn of the century, total fuels devoted to electricity generation were expected to exceed 100 quads per year—roughly 33 percent greater than the current total energy use in the United States and five times greater than the present rate of electricity usage.

Figure 2-1.   Projections of Total Energy Demand (1978–2010).

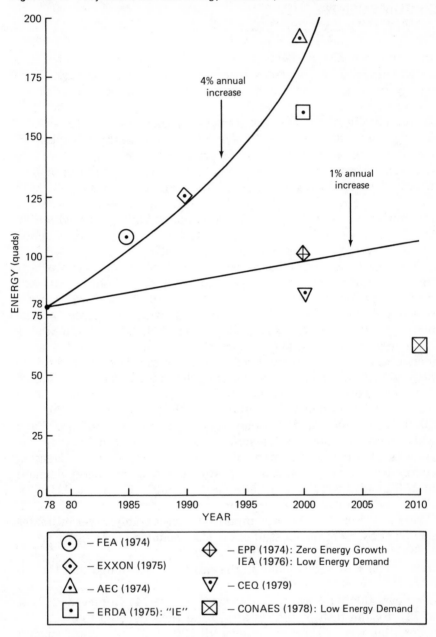

To simplify the energy accounting, we have adopted the "primary energy equivalent" convention throughout this study. All electrical energy values, independent of the manner in which the electricity is produced, are multiplied by three to obtain a primary energy equivalent or gross thermal energy value. This convention is used to normalize electrical energy values from a variety of sources having different generation efficiencies. The factor of three arises from the fact that present conventional electric power plants are about 33 percent efficient, meaning that three units of thermal energy input are required to produce one unit of electrical output.

Since electricity use has grown substantially faster than gross energy consumption, a curious difficulty can arise in forecasts based solely on historical trends. For example, one report prepared in 1971 for the U.S. Department of Commerce (1971) concluded that electricity use would grow uniformly at 5.46 percent per year from 1970 to 2040, increasing overall by a factor of forty. Over the same period, a lower growth rate of about 2.55 percent was predicted for total energy use, leading to the conclusion that "by about 2015–2025, the amount of energy channelled into electricity will be *greater than* the projected total energy usage - an obvious paradox" (p. 8, original emphasis). The report noted that extrapolations extended over long periods and could not be entirely reliable, but concluded that "nevertheless, in the absence of better information, we have no choice but to extrapolate using known factors" (p. 8). No attempts were made to resolve the paradox by a detailed study of the end uses for energy and electricity.

Somewhat lower growth rates for electricity and energy in general were made following the oil embargo of 1973–1974 in response to increased fuel prices and expectations of slower GNP growth. Projections for growth in the electricity sector were reduced dramatically. For example, 1973 government forecasts for installed nuclear capacity (about 50 gigawatts [GWe] early in 1979) in the year 2000 reached a high value of 1500 GWe. Projections made in late 1978 had dropped off as low as 157 GWe (DOE 1978). The sharp decline in projections for installed nuclear capacity reflected not only substantially lowered assumptions of electricity growth, but also mounting problems confronting the nuclear industry. Nevertheless, it is clear that estimates that shift so dramatically in such a short period of time rest on very fragile foundations.

Regardless of the actual numbers arrived at, all projections of this nature completely ignore how the increased supply of energy will ultimately be used. If, for example, a 4 percent growth in energy use is assumed, per capita energy use will more than double between now and the year 2000, whereas it has increased only threefold over the past 125 years, despite the heavy industrialization that has taken place. As will be discussed later in the chapter, an average increase in per capita energy use of only about 50 percent would allow everyone in the United States to use energy at a level corresponding to that of the highest (upper 20 percentile) income group in the nation today. It is far from obvious how energy could be productively (or even wastefully) put to use at the much higher levels implied by conventional forecasts.

Projections for electrical growth typically fail to incorporate the appropriate role that should be assumed by electricity among the spectrum of useful energy forms. The use of electricity grew at a nearly constant level of 7 percent per year from 1900–1976. While spectacular growth in the use of electricity was surely warranted in previous decades when electricity was first being introduced to new markets, saturation levels for electric appliances are already high throughout the country (EPP 1974), and the use of electricity has already been extended beyond the range of functions for which it is most suitable. Therefore, projections that assume continued growth at rates near 7 percent per year have little basis in reality.

For example, in the "intensive electrification" scenario devised by the Energy Research and Development Administration (ERDA 1975), the use of electricity for industrial process heat was expected to exceed 22 quads (primary energy equivalent) per year by 2000. No economic rational is (or can be) provided for this assumption. In fact, virtually no use of electricity is made for this purpose today because industries have long realized that it is uneconomical (in addition to being thermodynamically inelegant) to do so. It is exceedingly unlikely that circumstances arising in the future will change in such a way as to make electricity a prudent means of supplying process heat.

The absurdity of making long-term predictions on the basis of simple extrapolations is summed up well in the following quote:

> In the space of one hundred and seventy-six years, the Lower Mississippi has shortened itself two hundred and forty-two miles. That is an average of a trifle over one mile and a third per year. Therefore, any calm person, who is not blind or idiotic, can see that in the Old Oolitic Silurian Period, just a million years ago next November, the Lower Mississippi River was upward of one million three hundred thousand miles long, and stuck out over the Gulf of Mexico like a fishing rod. And by the same token any person can see that seven hundred and forty-two years from now, the Lower Mississippi will be only a mile and three-quarters long. . . . (Twain 1961:120)

ESTIMATING FUTURE ENERGY REQUIREMENTS: AN ALTERNATIVE APPROACH

Our estimates of future energy requirements are based on a wholly different, although admittedly simple, approach. Predictions extending far into the future are necessarily uncertain. It is not possible to estimate long-term energy requirements to a high degree of precision; any attempts to do so are simply misleading. The best that can be done is to suggest the general magnitude of future energy demand under an explicitly stated set of assumptions. Rather than employing the conventional technique of extrapolating from historical trends, our estimates are built on modifications of the existing structure of energy use. The year 1975 was selected as the base case. First, patterns of energy use in the residential, commercial, industrial, and transportation sectors in 1975 are examined, and estimates are made regarding the amount of energy that could have been saved had the energy been used more efficiently. Next, estimates are made for population growth, also a determinant of future energy use, and the impacts of hypothetical improvements in the standard of living on future energy demand are assessed.

Given the prevailing uncertainties, it is preferable to come up with a range of possible values for future energy requirements rather than a single point. Thus a range of values has been assumed for each of the relevant parameters—energy efficiency, population growth, and per capita energy use—in order to incorporate the combined effects of the individual uncertainties into estimates of aggregate demand.

This procedure yields a range of estimates for future energy needs that are far lower than those arrived at by means of conventional extrapolations of exponential energy growth. In fact, our analysis indicates that with the implementation of numerous technical measures designed to improve the efficiency of energy utilization, long-term energy requirements for the United States would not greatly exceed and may even be lower than present levels, even if generous increases in effective per capita energy use—reflecting elevated living standards—are assumed. Furthermore, the adoption of the many efficiency measures required to curb energy growth need not induce economic stagnation. On the contrary, available evidence suggests that the overall effects on the economy should be decidedly positive, because up to a certain point (which we are still far from approaching), it is far cheaper to reduce energy demand through improved efficiency than to increase energy supplies to a corresponding degree.

In our future scenarios, we have assumed that a transition from our present energy system—allowing for the extensive implementation of energy-efficiency improvements, as well as for the deployment of major new energy supply technologies—could be achieved by the year 2050. Although this date is admittedly arbitrary, it does allow a reasonable period in which the United States could achieve a stable population and a correspondingly stable level of energy consumption. This time frame could, of course, be compressed or expanded, depending on a number of contingencies and changing imperatives. Energy requirements have also been estimated for the year 2000, as well as for 2050, in order to provide a more immediate guide for policy decisions to be made in the near future.

PRESENT PATTERNS OF ENERGY USE

Energy use within the individual consuming sectors—residential, commercial, industrial, and transportation—along with the rough percentages of gross energy used in the form of electricity for the year 1975 are presented in Table 2-1 based on data compiled by the Department of the Interior (1976) and Ross and Williams (1975).

Energy use by sector for the year 1978 is presented in Table 2-2. From this table it can be seen that while total energy use increased by about 10 percent from 1975 to 1978, the apportionment of energy into the various consuming sectors remained essentially unchanged.

The breakdown of 1978 energy consumption according to end use category is presented in Table 2-3. Included under the heading of "Net End Use Energy" is the energy delivered to the consumer in the form of oil, gas, coal, gasoline, electricity, and so forth. Approximately 71 percent of the primary energy devoted to electricity pro-

Table 2-1. Energy Use by Sector for 1975—Baseline Data.

Sector	Energy (in quads)	Percent of Total	Percent Electrical (primary energy equivalent)
Residential	13.6	19.1	45.0
Commercial	11.6	16.3	45.0
Industrial	27.3	38.4	30.0
Transportation	18.6	26.2	—
Total	71.1	100.0	

Source: Based on data compiled by the Department of the Interior (1976) and Ross and Williams (1975).

Table 2-2. Energy Use by Sector—1978.

	Energy (in quads)		
Sector	*(w/o electricity)*	*(with electricity[a])*	*Percent of Total*
Residential and Commercial	14.4	28.6	36.6
Industrial	19.5	29.0	37.1
Transportation	20.5	20.6	26.3
Total	54.4	78.2	100.0

[a]Primary energy devoted to electrical generation, including losses incurred in generation and transmission.

Source: DOE (1979a).

duction in 1978 was lost in conversion and distribution (DOE 1979a) and is therefore not considered "end use" energy. It should be noted that additional conversion inefficiencies—not considered in this table—will result at the point of utilization. For example, energy will be lost when oil is burned in a furnace to heat a home, gasoline is consumed in an internal combustion engine to power an automobile, and electricity is converted to light in a typical incandescent light bulb.

THE POTENTIAL FOR ENERGY SAVINGS

As illustrated in Figure 2-2, the potential for energy savings in the U.S. economy is considerable: the overall "first law" efficiency (defined as the ratio of the net useful energy provided to the primary energy input) of energy use in the United States is only about 36 percent (Cook 1976), meaning that nearly two-thirds of the primary

Table 2-3. Energy Use in 1978 by End Use Form.

End Use Category	*Gross Primary Energy (in quads)*	*Percent of Gross Primary Energy*	*Net End Use Energy (in quads)*	*Percent of Net End Use Energy*
Thermal	28	36	28	46
Electrical	24	31	7	11
Transportation				
Fuels	21	27	21	34
Feedstocks	5	6	5	8
Total	78	100	61	99[a]

[a]Does not add up to 100 due to rounding.

Source: DOE (1979a, 1979b).

Figure 2-2. Energy Flow in the U.S. Economy in 1975. Food energy is excluded. All figures are in trillions of megajoules (10^{12} MJ = 10^{15} kJ = 0.95×10^{15} Btu). Hydropower is computed as 100 percent efficient. Changes in inventories (invariably relatively small) are lumped with net imports or exports. Industrial consumption includes nonfuel uses of energy resources (e.g., asphalt, petrochemicals). Efficiency figures used to disaggregate useful heat and work from waste are very rough estimates of first law efficiency: household-commercial, 72 percent; transportation, 15 percent; industrial, 40 percent. Electric generation efficiency averages about 38 percent when hydro is included.

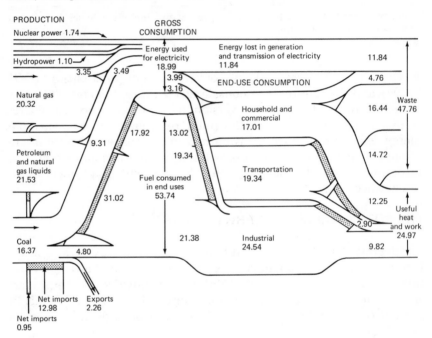

From *Ecoscience: Population, Resources, Environment* by Paul R. Ehrlich, Anne H. Ehrlich, and John P. Holdren. W.H. Freeman and Company. Copyright © 1977.

energy consumed is "wasted." For example, generating efficiencies at thermal-electric power plants average roughly 33 percent, implying that two-thirds of the primary energy input is discharged as low grade thermal energy or "waste" heat. While some of these conversion losses are necessary on fundamental thermodynamic grounds, overall efficiencies can nonetheless be substantially improved by making use of the rejected residual heat. Automobiles are even less efficient, typically converting only about 15 percent of the fuel energy into mechanical energy delivered to the wheels; energy lost through deformation and heating of the wheels further diminishes the overall efficiency (Cook 1976).

The overall potential for energy savings is obscured and may in fact be understated by considering only the "first law" efficiency. While the first law efficiency can tell us how well a specific device performs, it sheds almost no light on the most efficient approach to perform a particular task. The concept of the "second law" efficiency, which considers both the quantity and thermodynamic quality of energy used, provides a more useful yardstick for this purpose (APS 1975; Ross and Williams 1977). Simply defined, the second law efficiency is the ratio of the theoretical minimum amount of energy required to perform a designated task to the amount of energy actually used in performing the task. On the basis of the second law of thermodynamics, the overall efficiency of energy use in the United States has been estimated to be only about 8 percent (Widmer and Gyftopulous 1977). It would be impossible, of course, to achieve 100 percent efficiency in practice due to the limitations of technology (i.e., the impossibility of manufacturing a "perfect" device), time (i.e., the impracticality of carrying out reactions so slowly as to be completely "reversible"), and economics (i.e., at some point, additional capital investments will outweigh the value of incremental energy savings). Ross and Williams (1977) have suggested average second law efficiency values of twenty to fifty percent as being practicable, long-term goals—entailing roughly a three- to six-fold overall improvement in energy efficiency.

Energy saved through increased efficiency constitutes a significant energy "source." For example, if the overall efficiency of energy use in the United States were increased by only 10 percent, the amount of energy saved over an eight-year period would exceed the total energy expected to be recoverable from Alaska's North Slope oil fields. A 25 percent improvement in national energy efficiency could thereafter free up on an annual basis an amount of energy exceeding the fraction we currently obtain from imported oil. Therefore, possible improvements in the way we utilize energy clearly deserve the utmost consideration.

There are two basic strategies for improving the efficiency of energy use. One is to rely strictly on technical measures to reduce energy waste without any change in energy services. The second approach would involve changes in behavior and traditional consumption patterns designed to reduce the need for energy. Technical "fixes" can assume several general forms. We can make use of the energy that is currently wasted. Energy waste can in turn be reduced by more closely matching primary energy sources with end use tasks.

Measures can also be implemented to reduce the amount of energy needed to provide a given service or to produce a given good. This can be accomplished either by improving the energy efficiency of existing buildings, devices, factories, or vehicles or by replacing them with newer, more efficient models. Examples of technical "fixes" include utilizing the waste heat at thermal-electric power-generating stations; reducing building heat loss through increased insulation; increasing the efficiency of lights, motors, and electrical appliances; the use of heat recuperation devices in residential and commercial buildings and industries; the industrial "cogeneration" of steam and electricity (whereby the typically disparate tasks of steam and electricity production are combined); and improving the fuel economy of transportation vehicles by moving to lighter weight, streamlined designs and employing more efficient heat engines. Technical fixes involve not only improving the efficiency of a particular device—be it an automobile, furnace, water heater, air conditioner, or light bulb—but also shifts to less energy-intensive means of providing a particular service by substituting other inputs for energy. For example, while the amount of energy required to produce a given metal from virgin ores can be reduced considerably by employing more efficient production techniques, even greater energy savings can be obtained through the recycling of scrap metals. Thus, in order to maximize energy savings, it is necessary to identify the most energy-efficient method of fulfilling a specific need, rather than simply improving the efficiency of the conventional supply approach.

Beyond technical fixes, further energy savings could be achieved by social shifts toward less energy-demanding lifestyles. It is not at all necessary that such shifts be unpalatable or distasteful. In fact, many can enhance, rather than detract from, the quality of life. Changes of this nature could include shifts away from highly energy-intensive products, living and transportation arrangements, and recreational activities. For example, a greater reliance could be placed on walking, bicycling, public transportation, and carpooling instead of the private automobile. Many of these lifestyle changes would involve little if any personal sacrifice and even yield benefits in some cases. However, apart from assuming some general shifts in transportation modes, the emphasis in this analysis has been on technical fixes (partly because they are simpler to evaluate). Nonetheless, the fact remains that we can save considerable amounts of energy solely through technical improvements in energy efficiency. The advantages and disadvantages of saving additional amounts of energy through long-term shifts in lifestyles should be carefully weighed on a case-by-case basis.

UCS MODEL: ENERGY
EFFICIENCY CONSIDERATIONS

We have concluded from our analysis that efficiency improvements introduced by the year 2050 could reduce the average amount of energy consumed per unit good produced or service provided by 50 percent in comparison to present use patterns. A more effort. Given the long time frame involved, we believe these estimates to be quite conservative and easily within the technical capabilities of the United States. Indeed, all the efficiency improvements we contemplate are based on technology that is at hand today or will be available in the near future. For the most part, these efficiency techniques and devices are already economically cost-effective under present market conditions. Their economics will only improve as the price of conventional fuels continues to increase in the future. Although additional technical advances are almost certain to occur in the future, we have not assumed them in our long-range scenarios. Obviously, inefficiencies in energy use could be further trimmed with these advances.

A number of major energy efficiency studies support our assumptions regarding achievable improvements (EPP 1974; Ross and Williams 1976; Widmer and Gyftopolous 1977; APS 1975; AIA 1975). An important general conclusion of these studies is that significant reductions in energy use can be achieved by purely technical measures, without a lowering of the standard of living or resorting to major lifestyle changes. In fact, the net effect of any assumed lifestyle changes would be an overall enhancement of the quality of life due to reduced environmental degradation and an improvement of public services, such as mass transportation. The net economic impact of reducing energy demand by the implementation of energy-efficient technologies, rather than increasing energy supplies, would be overwhelmingly beneficial.

Studies by the Energy Policy Project of the Ford Foundation (EPP 1974) and Ross and Williams (1976) are especially noteworthy. The Ford Foundation study concluded that by the year 2000, strictly technical efficiency improvements could reduce overall energy use by 34 percent in comparison to a continuation of present practices. If minor lifestyle changes are introduced, energy use could be reduced by 46 percent. It is worth pointing out that the lifestyle changes incorporated in the latter scenario allowed for a net increase in energy use and travel on a per capita basis, although some shifts in transportation modes were called for. Ross and Williams examined energy use in 1973 and concluded that by the year 1990, overall energy use could be reduced by 42 percent, predicated solely on the basis of more efficient practices. The degree of conservatism in our 30–50 percent efficiency improvement estimates for the year 2050 comes to light when it is realized that the 34–46 percent reductions in gross energy use arrived at by the Ford Foundation study and the 42 percent energy savings derived by Ross and Williams are for the years 2000 and 1990, respectively.

The potential for energy savings in the United States can be illuminated by means of comparison to the industrialized countries of Europe, which use energy much more efficiently. For example, Sweden uses about 40 percent less energy than the United States on a per capita basis, but has a higher average per capita income (Schipper and Lichtenberg 1976). As stated by Schipper (1975): "The Swedes have more medical care, people-oriented services, and mass transportation, but fewer large cars, flimsy products, leaky houses, and empty buses" (p. 23). A study by Goen and White (1975) reached similar conclusions about energy use in West Germany, which has a per capita gross national product comparable to that of the United States, with less than half the per capita energy use. Conditions vary in different countries, and thus comparisons of this nature should be made carefully. Nonetheless, the present similarity in both per capita gross national product and industrial production in the United States, Sweden, and West Germany enhances the validity of these particular comparisons (Ross and Williams 1977).

We will now briefly discuss electricity, an energy form that fulfills a unique role in all energy use sectors. This will be followed by a review of the potential for energy efficiency gains in the residential and commercial, industrial, and transportation sectors, in order to provide validation for our general findings regarding possible efficiency improvements.

Electricity

Electricity is an intermediate form of energy made only from primary fuels or by the conversion of sunlight or wind and water power. Owing to its high thermodynamic quality and great flexibility of usage, electricity holds a unique place among the spectrum of energy forms. Many tasks can be performed only by energy of such high quality. However, the technology required to produce electricity makes it a very expensive energy form. Therefore the use of electricity is economically justified only when it is matched to a high grade task. Furthermore, it is also important to closely match the use of electricity with appropriate tasks on energy efficiency grounds, because it is generated very inefficiently in general by means of a process that is inherently wasteful of primary energy sources.

As indicated in Table 2-3, electricity production presently accounts for about 31 percent of the primary energy used in the United States. Examined on an end use basis, electricity currently comprises about 11 percent of all energy delivered to users, deducting that lost in conversion and distribution.

Future electrical requirements could be reduced if currently nonessential uses of electricity—such as electric space heating, water heating, air conditioning, cooking, clothes drying, and industrial process heat and steam applications—were eventually met instead by thermal energy. For example, residential and commercial space heating, water heating, and air conditioning together account for roughly 20 percent of the electricity consumed in the United States (Ross and Williams 1975). All of these tasks can be accomplished using thermal energy sources, although in some cases, especially for air conditioning, it might prove convenient to continue to use electricity. The intrinsic diseconomies involved in using electricity for space heating—an application often advocated by electric utility companies—is ironically demonstrated by the fact that most, if not all, of the nation's nuclear power plants are heated by oil-, gas-, and coal-burning unit heaters rather than by electric systems.

We have assumed that in the long term, the use of electricity in the residential and commercial sector will be limited to essential purposes such as lighting and the operation of electrical motors, appliances, and refrigerators. Similarly, for the industrial sector, it is assumed that electricity is used only for tasks (including electric drive, electrolytic, electrochemical, and other electric applications) specifically requiring its use, but not for process steam or direct heat applications. The adoption of these modifications in the residential, commercial, and industrial sectors would lead to thermal energy replacing roughly 25 percent of all the electricity presently used. Restricted to appropriate uses, electricity would currently comprise about 20 to 25 percent of overall primary energy usage and only about 8 percent of net end use energy requirements.

An important method of increasing the efficiency of electrical generation is through the employment of total energy systems, which utilize the waste heat from the generation of electricity in order to provide both electricity and usable thermal energy. Total energy concepts include district heating systems, industrial cogeneration, and a variety of decentralized systems. In district heating systems, low temperature thermal energy that is normally discharged to the environment as waste heat from electric generation is instead used to provide energy for residential and commercial space and water heating, industrial processes, the desalinization of water, and agricultural purposes. The overall efficiency of primary fuel utilization can be increased from a typical value of about 33 up to 85 percent (Karkheck et al. 1977). Industrial cogeneration is discussed later in this chapter. Fuel cells and solar thermal-electric systems might also be used in the future to provide heat and electricity. These systems will be discussed more fully in later chapters.

It seems unlikely that our approach of restricting electricity use to appropriate applications will lead to serious underestimates of future electricity requirements. While in some instances the use of electricity for tasks such as air conditioning, industrial process heat, cooking, and drying may prove advantageous on grounds of convenience rather than thermodynamics, this will be counteracted by other measures that will tend to reduce electricity demand. For example, although we have assumed that electricity will continue to be used for refrigeration, other energy forms will be suitable as well: gas absorption refrigeration units could be run off any intermediate temperature heat source, and even ice could be used to provide cooling in some cases. This consideration is not trivial given that refrigeration presently accounts for about 10 percent of all electricity use (Ross and Williams 1975). Similarly, lighting accounts for about 20 percent of the current total electricity consumption (Ross and Williams 1975). Available evidence indicates not only a great potential for technical efficiency improvements—perhaps by a factor of ten or more over incandescent and fluorescent light bulbs (Ayers and Kramer 1976)—but also shows that considerable electricity is wasted owing to unnecessarily high illumination levels (Hayes 1977). In addition, a reliance on total energy systems (including district heating and industrial cogeneration) can further reduce the amount of primary fuels devoted exclusively to the generation of electricity. The percentage of gross energy used in the form of electricity may change to some extent if higher per capita energy use levels are attained resulting in a greater use of electrical appliances, for example. However, because saturation levels for most major appliances are nearly achieved, only minor increases in electricity consumption would occur (EPP 1974).

Residential and Commercial Sectors

Our estimate of long-term savings of 30–50 percent in the residential and commercial sectors is certainly consistent with estimates made within the general literature. The Ford Foundation study group projected possible energy savings of 44 percent in the residential sector (EPP 1974). Ross and Williams (1976) identified measures that would reduce energy consumption by 47 and 41 percent in the residential and commercial sectors, respectively. A study by the American Institute of Architects (1975) concluded that the adoption of improved construction techniques could reduce by 60 percent the total energy requirements of new residential and commercial buildings constructed through 1985, whereas energy savings of 30 percent were assumed achievable by retrofitting existing buildings for energy efficiency. In both cases, dollar savings would be realized along with energy savings. Bliss (1976) found that the heating requirements of the average U.S. home could easily and economically be cut in half. He further concluded that custom-designed homes could be built with heating demands that were only 20 percent of those of a "normal" home. Similarly, a study by the American Physical Society (APS 1975) outlined simple measures that could reduce the heating demands of the 'standard' house by a factor of almost four. Ross and Williams (1977) also found that home heating losses could be reduced by almost 75 percent by straight forward innovations in design. They note the significance of heating load reductions of this magnitude by pointing out that virtually all home heating needs could be provided by sunlight entering through windows and by the heat internally generated by appliances and people, except on extremely cold days.

The need for electricity and natural gas in buildings was reduced by an average of 36 and 61 percent, respectively, in an Ohio State University program (Nader 1976). These conservation measures paid for themselves as a result of energy savings within four and one-half to thirteen months. In a conservation project in Princeton, New Jersey, a townhouse was economically retrofitted with insulation, reducing the demand for space heat by 67 percent. Houses presently built in Sweden use only one-half as much energy per unit floor area and degree day as those constructed in the United States (Schipper and Lichtenberg 1976).

The basic strategy for improving energy efficiency in the residential and commercial sectors is relatively straightforward. First, the design and construction of buildings can be improved to reduce heating and cooling loads. This can be accomplished by a combination of measures including improved insulation, better windows and more thoughtful window placement, and decreased ventilation by means of general "leak plugging," such as caulking and the installation of weather stripping. Second, significant improvements can be made in the efficiency of heating systems, air conditioners, water heaters, lights, and major appliances such as refrigerators, and devices such as heat recuperators can be installed. For example, supplies of natural gas can be stretched simply by the elimination of wasteful pilot lights. Similarly, major savings of energy can result from the replacement of electric resistance heaters with more efficient heat pumps.

Industrial Sector

Achieving long-term efficiency gains of 30–50 percent appears to be a reasonable expectation in light of other studies and the existing evidence. Studies by the Ford Foundation (EPP 1974) and Ross and Williams (1976) estimated potential savings within industry of 35–51 percent and 35 percent, respectively.

A comparison with various industries in Europe is also instructive. For example, West German industries require 42 percent less energy to manufacture a ton of paper and 32 percent less energy to produce a ton of steel than U.S. industries (*Washington Post* 1977). The nation's steel industry currently uses over 4 percent of all the energy used in the country. By employing technology currently available in the United States, the amount of energy required to produce a ton of steel can be reduced by 30 percent (*Steel* 1978). Similarly, American chemical companies use 20 percent more energy to produce a ton of polyvinyl chloride, a widely used plastic material, than companies in Britain, Italy, and the Netherlands. Schipper and Lichtenberg (1976) concluded that Swedish industry was about ten to fifteen years ahead of the United States on the basis of energy efficiency. All of this is not to suggest that the relatively higher efficiency of European industry leaves no room for further advances. For example, Swedish researchers have identified a process that would cut in half the amount of energy used in steelmaking in Swedish industries—industries which are already well ahead of their U.S. counterparts (Eketorp 1978).

Potential energy savings are more difficult to estimate in the industrial sector than in the residential and commercial sectors owing to the variety and complexity of the numerous activities involved. Nonetheless, several important measures can be taken to improve the energy efficiency of industry. These include the replacement or retrofitting of inefficient industrial facilities and equipment; the establishment of more efficient production processes; a general shift to less energy-intensive production methods; the use of heat recuperators and regenerators; the on-site cogeneration of steam and electricity; and the recycling of urban and industrial refuse for materials extraction, fuel production, direct combustion, and electricity generation.

The recycling of scrap metal, glass, paper, and other materials is highly sensible from an economic, environmental, and energy efficiency point of view. According to one estimate, the extraction and processing of raw materials presently accounts for about two-thirds of all U.S. industrial energy use or, equivalently, one-quarter of all U.S. energy use (Hayes 1977). The amount of energy required to produce some important products such as metals, glass, and paper from urban and industrial waste may be far less than that required if starting from raw materials. For example, the manufacture of aluminum is one of the most energy-intensive industrial activities in the United States, requiring an abundant supply of inexpensive electricity. Ralph Nader has referred to aluminum as "congealed electricity." However, only 5 percent as much energy is required to recycle aluminum (including separation, transportation, and processing) as is needed to produce it from ore; for copper and steel, the respective figures are 9 and 14 percent, respectively (Hayes 1977).

Industrial cogeneration offers an even greater potential for saving energy. The conventional means of producing steam for industrial process applications involve the burning of a high quality fuel to boil water. This practice is thermodynamically wasteful because the high quality energy of the fuel is not fully utilized in performing a task normally requiring relatively low temperature energy. A more efficient method is first to use the fuel for the generation of electricity and then to utilize the steam rejected from the turbine (with its burden of heat normally considered waste) for industrial process applications. This technique is referred to as cogeneration because two useful forms of energy—steam and electricity—are generated simultaneously. The amount of extra fuel required to produce electricity under this arrangement is only about one-half that which would have been required to produce the electricity separately (Dow Chemical 1975).

Industrial process steam generation accounts for approximately 15 percent of total energy use in this country, so the national potential for cogeneration is quite large. A report by Dow Chemical Company (1975) estimated that by 1985, 71,000 MWe of cogeneration capacity could be installed—enough to provide about 50 percent of the present electrical needs of U.S. industry. This figure is about 40 percent greater than the entire 1978 installed nuclear capacity in the United States. The Dow study also found that, in addition, the production of industrial process steam through the utilization of the waste heat of central power plants could result in energy savings of 1.1 quads in 1985. A study by the Thermo Electron Corporation was even more optimistic. It concluded that roughly 135,000 MWe of cogeneration capacity could be installed by 1985 (Nydick et al. 1976)—equivalent to about 25 percent of the total electrical generating capacity existing in the United States. Ross and Williams (1975) estimated that this level of industrial cogeneration could lead to annual energy savings of 2.6 quads. Looking to the year 2000, Williams (1978) concluded that installed cogeneration capacity could displace more than 200,000 MWe of conventional electrical generation capacity (equivalent to about 200 nuclear plants) and, in so doing, save 5 quads of primary energy annually.

Chemical Feedstocks

Feedstocks are fuels used as raw materials rather than as sources of energy. Petrochemical industries—responsible for the production of chemical fertilizers, plastics, rubbers, lubricants, organic compounds, and a variety of synthetic materials—are a subset of the industrial sector dependent upon carbon feedstocks. Altogether, oil and natural gas and, to a lesser extent, coal consumed as raw materials for road paving and the production of fertilizers and petrochemical products add up to about 6 percent of total energy use in the United States. With the eventual depletion of oil and natural gas resources, feedstocks to these chemical industries will become of increasingly critical importance.

The total energy currently used in petrochemical industries is estimated to be about 5 quads, (EPP 1974b; Ross and Williams 1975). In order to determine future energy requirements, energy presently used for process heat—accounting for 15 percent of petrochemical energy use—can be subtracted from the total, since this energy can be provided by other sources. Some fraction of the 1.1 quads consumed in the form of asphalt and road oil can be reduced by the use of other construction materials, such as concrete. Further reductions in long-term energy requirements can be achieved by efficiency improvements in manufacturing processes and by means of substitution and recycling.

Transportation

At present, transportation holds a dominant position among energy use sectors, accounting directly for about 25 percent of the total energy use and for more than 40 percent when indirect costs such as highway construction and automobile manufacturing and maintenance are included (Hayes 1977). Automobiles and trucks respectively account for approximately 53 and 22 percent of total energy use in the sector. Air transport comprises about 8 percent, and the remaining 17 percent is used by agricultural vehicles, water vessels, pipelines, rail transport, buses, and mass transit systems. A breakdown of transportation energy demands is summarized in Table

Table 2-4. Distribution of Transportation Energy Requirements.

Transportation Mode	*Approximate Percentage of Total Transportation Energy*
Automobile	53
Truck	22
Air	8
Agriculture	5
Water	4
Pipeline	4
Railroad	3
Bus	1
Mass Transit (Electric)	0.05
Total	100

Source: Department of Transportation (1975).[b]

2-4 based on data from the Department of Transportation (1975).

The decline of the petroleum era will necessitate profound changes in future transportation systems, for petroleum supplies about 97 percent of the energy currently used in the sector. Energy supply options for transportation in the long term are assessed in Chapter 6. Even though it is impossible to predict the precise structure of a future transportation sector, a rough estimate of future energy requirements can be obtained by evaluating the impacts of measures designed to improve the efficiency of present transportation vehicles and simultaneously considering the effects of shifts in reliance from one mode of transportation to other more efficient methods.

The Ford Foundation study found that efficiency improvements and mode shifts could reduce transportation energy requirements by 37 to 55 percent (EPP 1974). Ross and Williams (1976) concluded that efficiency measures could reduce overall energy needs for transportation by 48 percent. Both of these studies support our conclusion that energy savings of 30-50 percent are possible by 2050. For a number of reasons to be discussed subsequently, per capita energy use for transportation in European countries is only about one-fourth that of the United States (Ross and Williams 1977). Specific areas where significant savings can be made will now be discussed.

General efficiency improvements of varying degrees can be made in all transportation vehicles. Because automobiles presently dominate energy use in transportation, major energy savings can be realized by increasing their efficiency. A study by the American Physical Society (APS 1975) proposed modifications in automobiles that could reduce gasoline consumption by about 60 percent. These modifications were considered achievable for a vehicle in the 1980s and included a reduction of average weight and engine power, improved transmission, reduced aerodynamic drag through body redesign, reduced rolling resistance through improved tires, and a switch to lightweight diesel or other improved engines. Similarly, energy savings can also be achieved by designing lighter trucks and buses and by switching from the relatively inefficient gasoline engines to diesel and other advanced engines.

Overall efficiencies for electric-powered vehicles, including generation and transmission losses, are roughly comparable to those of fuel-powered vehicles. However, the efficiencies of electric vehicles may be improved by means that are not available for conventional fuel-driven vehicles. For example, "regenerative" braking systems can recover most of the kinetic energy normally converted to heat during braking (NAS 1976). Regenerative braking devices have been installed on electric postal vans at modest costs. Flywheel storage systems, installed to a limited extent on some of the nation's subways, can also be employed to recover kinetic energy typically dissipated by braking. Electrical transportation systems also offer the possibility of waste heat utilization—an option that is largely impractical for fuel-powered vehicles. Since diesel and electric-powered trains have comparable overall efficiencies (Rice 1974), we have assumed that all rail transportation will be electrified in the distant future, in light of the imminent scarcity of liquid fossil fuels.

Significant savings of energy can be achieved by increasing the passenger load factor—meaning the percentage of available seats actually filled—of our automobiles, airplanes, buses, and other vehicles. This is considered the single most important measure for improving the energy efficiency of air travel (EPP 1974). Energy use per passenger mile in European countries is only about one-half that of the United States, owing to higher vehicle occupancy and better gas mileage (Ross and Williams 1977). Reduction of travel speeds to levels optimal from an energy efficiency standpoint would also lead to energy saving, although of a lesser magnitude. Resultant increases in travel time would generally be minor, but warrant consideration.

Energy use by automobiles could be reduced through a greater reliance on buses, trains, and other forms of mass transit. Passenger travel by rail is twice as efficient as air travel (EPP 1974). The shifting of passenger traffic from air to high speed trains for short distance travel could save energy without significantly affecting travel time.

Freight transport by rail is about four times more efficient than by truck and far more efficient than air transport (EPP 1974). Therefore, major energy savings could be achieved by moving some fraction of truck and air freight traffic by rail. Electric rail transport and water freight systems have comparable energy efficiencies (Hirst 1973). Critical liquid fuel demands could thus be alleviated to some extent by shifting water freight to electric rail.

Although not incorporated into our estimates, further reductions in travel requirements could be accomplished by changes in living patterns. For example, communities of the future could be designed in such a way as to minimize travel distances to places of work and to shopping and other service centers. Mass transit systems would become an integral part of urban design. In fact, the average distance traveled per year by a person in Germany and Sweden is only one-half that for the United States, partly attributable to the fact that average commuting distances are shorter in the more compact European metropolitan areas (Ross and Williams 1977). Public transportation is much more important in Germany and Sweden, and walking and bicycling are relied on in place of driving for short distances. Travel requirements in the future might be further reduced through a reliance on video telecommunication systems, which are in limited use today. We have not assumed these changes, but the advantages stemming from their implementation—in terms of reduced energy use and elimination of routine and unnecessary travel—make them worthy of serious consideration.

UCS MODEL: POPULATION GROWTH AND STANDARD OF LIVING

By taking into account potential efficiency improvements, it is possible to determine the average per capita energy requirements needed to maintain the 1975 standard of living if energy were used more efficiently. Total energy use is a function of both total population and average per capita use. Therefore, in order to evaluate future energy requirements, one must consider the influence of population growth and possible increases in average per capita energy use reflecting improved living standards.

The future population of the United States is, of course, subject to uncertainty. Most projections for the U.S. population assume an average growth rate (birth rate minus death rate) of about 0.5 percent per year or less. A more illuminating demographic statistic is the fertility rate, which takes into account differences in the sex ratio and age composition of the overall population. The total fertility rate, measuring the average number of lifetime births per woman, is the most important determinant of population over the long run. Between 1973 and 1976, the total fertility rate maintained a value between 1.8 and 1.9, which is below the replacement level of 2.1 (Ehrlich, Ehrlich, and Holdren 1977). If this trend were continued and net immigration were maintained at its present rate of 400,000 per year, population growth would stop around the year 2025, with the total population at a level of about 250 million, gradually decreasing to about 240 million by the year 2050 (Bureau of the Census 1975).

At present, it is impossible to state whether the low fertility rates of the early 1970s will be sustained. Therefore, in order to bracket the uncertainties, we have selected two population growth scenarios prepared by the Bureau of the Census (1976). The lower estimate assumes that net immigration will continue at the present rate and that a total fertility rate of 1.7, slightly lower than the present value, will be maintained through 2050. With these assumptions, zero population growth would be achieved by the year 2020. The population in that year would be slightly over 240 million and would gradually decline to about 230 million by 2050. The higher growth estimate assumes a continuation of immigration at the present rate and a total fertility rate maintained at the replacement level of 2.1. In this case the population would increase to about 315 million by 2050. The low and high growth scenarios for the year 2050 represent net increases of 8 and 48 percent respectively over the 1975 U.S. population of 213 million.

It is interesting to note that even if the fertility rate was held at the replacement level, as in the latter scenario, zero population growth would not be achieved until 2050. This is due to the relatively young age structure of the present U.S. population. An immediate cessation of growth would require either a total fertility rate of 1.0 with present immigration rates or a rate of 1.2 with no net immigration.

By stabilizing energy and population growth, long-term energy demands can be maintained at a comparatively low level. In a rapidly expanding economy, such as has existed over the last several decades, much energy is expended simply for the development of new energy sources and the construction of new power plants. However, in a stable energy state, the bulk of these transient energy demands would disappear, leading to a relatively constant level of demand.

Although it is uncertain whether population growth will continue at a rate above or below current levels or whether in fact zero growth will ultimately occur, it is clear that the achievement of zero population growth warrants consideration as a national goal. By curbing population growth at an early date, the total population can stabilize at a relatively low level, thereby alleviating energy requirements, environmental pressures, and resource demands.

Per capita energy use is not uniform throughout society, but in fact closely parallels income level. Findings from a study of the relationships between energy use per family and socioeconomic factors are presented in Table 2-5. Included in the table are the average 1972 family incomes for four major groups and the ratio of average family energy use (including both direct and indirect use) for each income group relative to the upper income level. The average household was assumed to consist of 3.2 persons, and thus differences in energy use between families having different incomes can be readily computed on a per capita basis. Although average income levels have increased since 1972, relative values have not changed significantly. A similar relationship between energy use and income level was found by Herendeen (1974).

We wished to consider different distributions of energy use, in addition to the present one, that incorporate higher levels of effective use in order to evaluate the net effect of improving the "energy standard of living" on future energy requirements: (1) the present energy use distribution remains the same; (2) the poor and lower middle income group energy use levels are increased to that of the upper middle income level, while the upper income use rate remains unchanged; and (3) everyone uses the amount of energy that is used by the upper income group today. We will refer to these three cases as the current, intermediate, and high energy use levels. The last two cases represent an increase in average per capita energy use of 28 and 55 percent respectively over the first case. It is a remarkable fact that an increase in per capita energy use of only 55 percent would allow everyone in the United States to use as much energy as the highest income group does today. This is so especially in light of the fact that some projections of future energy demand (which we believe to be unrealistic) envision per capita increases exceeding 100 percent by

Table 2-5. Energy Use and Income Level.

Income Level	Percentage of Households	Average Family Income (1972)	Relative Energy Use
Poor	18	$ 2,500	0.36
Lower Middle	42	$ 8,000	0.54
Upper Middle	19	$14,000	0.78
Upper	20	$24,500	1.00

Source: EPP (1974).

the year 2000. Our high energy use case represents a level of energy comfort that can only be described as luxurious. There are some reasons why the United States may not wish to make achieving it a high priority. Yet, as we shall see, it certainly could be attained without the need for a major expansion of energy supplies.

In this study, we have made no attempt either to describe how such changes in per capita energy distribution could be implemented nor to evaluate their overall desirability. We wished only to determine how much energy would be required if they actually occurred. Such models enable one to relate a specific energy use level roughly to a given lifestyle in today's society.

USC MODEL: TRANSITIONAL ENERGY NEEDS

Our estimates of long-term energy efficiency improvements, population growth, and per capita energy use are important to the evaluation of the nation's energy future after the depletion of most fossil fuel resources and when wholly new energy sources have been developed and implemented in substitute. But we must additionally provide a near-term focus and make estimates for the "transition" period to the year 2000. This transition is of great importance; if it is poorly conceived and administered, then the hopes for satisfactory long-term solutions are at risk.

Estimates of transitional energy needs are closely related to the long-term analysis. The efficiencies assumed must be compatible with the long-term outlook, as well as realistic. Due to the long lead times necessary to fully implement some new technologies, it may not be possible to achieve all of the postulated long-term savings by the year 2000. For example, lead times in the residential housing sector are quite long, since buildings may last well over fifty years. Although the energy efficiency of existing buildings can be improved, the greatest potential for energy savings exists for new buildings. Building turnover times in the commercial sector are similarly long. Equipment turnover times in the industrial sector are shorter, because of the need for the continued modification and updating of manufacturing processes. The turnover times are shortest for the transportation sector—approximately ten to fifteen years for automobiles, trucks, buses, and airplanes. Railroads and water vessels have longer useful lifetimes, but presently account for only a small fraction of the total energy used for transportation. Because of the generally short inventory times typical in this sector, the projected gains here should be relatively easy to achieve.

To be conservative, we have assumed that one-half of the projected long-term efficiency improvements could be made by the year 2000, corresponding to energy savings of 15 to 25 percent. This does not mean that efficiency gains in new buildings, industrial processes, and transportation vehicles need only be one-half as large for the year 2050. On the contrary, over the next few decades these gains must be comparable to the long-term potential, in order to offset the relatively low efficiencies of existing energy conversion and end use facilities, which will still remain in part by the year 2000.

The same set of population growth rates has been assumed as for the long-term case. Population levels for the two scenarios will reach 245 and 262 million in the year 2000, representing increases over 1975 of 15 and 23 percent, respectively. As with the long-term projections, the effects of increased per capita energy use were examined for the year 2000, assuming the same set of increases.

UCS MODEL: ESTIMATES OF FUTURE
ENERGY REQUIREMENTS

Before computing gross energy requirements, it will be instructive to consider—at least in a rough way—how energy may be distributed among the various end use categories. As discussed previously, we have assumed that in the long term, currently inappropriate uses of electricity in the residential, commercial, and industrial sectors will be discontinued in favor of thermal energy, with electricity consumption largely restricted to those application for which it is uniquely suited. Under these conditions, thermal energy applications will account for roughly half of total energy demand, while demand for electricity in these three sectors will respectively amount to about 22, 38, and 32 percent of total primary energy requirements.[1] We consider that the use of electricity in the transportation sector, however, may rise dramatically for reasons that are discussed in Chapter 6. Electricity may therefore amount to 30 to 40 percent of gross primary energy requirements. Overall demand for electricity will of course be substantially lower if other energy sources, such as coal-derived synthetic fuels, are instead relied upon for transportation. Aggregate demand for liquid transport fuels, methane, and chemical feedstocks may comprise approximately 10 to 20 percent of the total, dependent upon

1. Although the current ratio of electricity to total gross energy in the industrial sector was placed at about 30 percent, rather than 32 percent, these modifications would in fact lead to a net reduction in electricity use. The reason for this discrepancy is that the previously cited figure for the industrial sector included the wholly nonelectrical feedstock component that has been disaggregated from the latter figure.

the role ultimately assumed by electricity in transportation. Rough estimates of the percentage distribution of future energy requirements into the different end use categories for the various consuming sectors are summarized in Table 2-6.

For the year 2000, we have assumed that the present ratios of electric to thermal energy use of about 45, 45, and 30 percent, respectively, are maintained in the residential, commercial, and industrial sectors.

Table 2-6. Approximate End Use Distribution of Long-term Energy Requirements.

	Percentage of Total Primary Energy Use
Thermal Energy	
Low Temperature (less than about 212°F)	
Residential and commercial space heating, water heating, and air conditioning	20-25
Intermediate Temperature 212-572°F)[a]	
Residential and commercial cooking and drying; industrial process steam and heat[b]	10
High Temperature (greater than 572°F)[a]	
Industrial process steam and heat	15-20
Total Thermal Energy	50
Electricity	
Residential and commercial lighting, appliances, and refrigeration	10
Industrial—electric drive, electrolytic, electrochemical, and other electric applications	10
Transportation	10-20[c]
Total Electricity	30-40[d]
Liquid Transport Fuels and Chemical Feedstocks	10-20[e]
Total	100

[a]Some fraction of this thermal energy requirement may be displaced by methane or thermally generated hydrogen.

[b]Assumes 30 percent of all industrial process thermal energy is used at temperatures below 572°F (Metz and Hammond 1978).

[c]The use of hydrogen as a transport fuel may displace some fraction of the electricity in this category.

[d]Electricity's percentage of the total has been represented on a primary energy equivalent basis.

[e]The range within this category primarily reflects the uncertainty over the extent to which electricity and hydrogen will be able to displace liquid fuels now providing the dominant portion of energy used for transportation.

In the transportation sector, however, it has been assumed that electricity will account for about 25 percent of the primary energy consumed, with the remaining 75 percent of the energy required used in the form of liquid fuels. This mix could be achieved by using electricity to supply roughly 25 percent of the energy used by automobiles, trucks, and buses and 75 percent of the energy used by trains. Small shifts on the order of 10 percent to electric rail and mass transit from automobiles, trucks, airplanes, and water vessels have also been assumed. Overall demand for electricity in the year 2000 will be lower if electricity is used less extensively for transportation.

The principal estimates incorporated into our energy needs computations are summarized in Table 2-7. A range of possible energy requirements—considering all possible combinations of the parameters embodied in Table 2-7—for the years 2000 and 2050, respectively, are present in Tables 2-8 and 2-9. Upper case estimates for electricity demand (on a primary energy equivalent basis) have been placed in parentheses. These estimates assume that electricity provides one-fourth and two-thirds of the energy used in transportation

Table 2-7. Summary of Estimates Used in Future Energy Needs Computations.

Parameter	Range of Values	1975–2000	1975–2050
Efficiency Improvements[a]	Low	15 percent reduction in gross energy requirements	30 percent reduction in gross energy requirements
	High	25 percent reduction in gross energy requirements	50 percent reduction in gross energy requirements
Population Growth	Low	15 percent net increase	8 percent net increase
	High	23 percent net increase	48 percent net increase
Per Capita Energy Use	Current	0 percent net increase	0 percent net increase
	Intermediate	28 percent net increase	28 percent net increase
	High	55 percent net increase	55 percent net increase

[a]Percentage reduction in gross energy requirements per unit of service.

Table 2-8. Energy Requirements for 2000 (in quads).[a]

Efficiency	Population Growth	Per Capita Energy Use	Residential	Commercial	Industrial	Transportation	Total
Low	Low	Current	13.3 (6.0)	11.3 (5.1)	26.7 (8.0)	18.2 (4.5)	69.5 (23.6)
	High		14.2 (6.4)	12.1 (5.4)	28.5 (8.5)	19.4 (4.8)	74.2 (25.1)
High	Low		11.7 (5.3)	10.0 (4.5)	23.5 (7.0)	16.0 (4.0)	61.2 (20.8)
	High		12.5 (5.6)	10.7 (4.8)	25.2 (7.6)	17.2 (4.3)	65.6 (22.3)
Low	Low	Intermediate	17.0 (7.6)	14.5 (6.5)	34.1 (10.2)	23.2 (5.8)	88.8 (30.1)
	High		18.2 (8.2)	15.5 (7.0)	36.5 (10.9)	24.9 (6.2)	95.1 (32.3)
High	Low		15.0 (6.7)	12.8 (5.8)	30.1 (9.0)	20.5 (5.1)	78.4 (26.6)
	High		16.1 (7.2)	13.7 (6.2)	32.2 (9.7)	22.0 (5.5)	84.0 (28.6)
Low	Low	High	20.6 (9.3)	17.6 (7.9)	41.4 (12.4)	28.2 (7.0)	107.8 (36.6)
	High		22.0 (9.9)	18.8 (8.5)	44.2 (13.3)	30.1 (7.5)	115.1 (39.2)
High	Low		18.2 (8.2)	15.5 (7.0)	36.5 (10.9)	24.9 (6.2)	95.1 (32.3)
	High		19.4 (8.7)	16.6 (7.5)	39.0 (11.7)	26.6 (6.6)	101.6 (34.5)

[a]Estimated values for electricity demand (on a primary energy equivalent basis) have been placed in parentheses.

Table 2-9. Energy Requirements for 2050 (in quads).[a]

Efficiency	Population Growth	Per Capita Energy Use	Residential		Commercial		Industrial		Transportation		Total	
Low	Low	Current	10.3	(2.3)	8.8	(3.3)	20.6	(6.5)	14.1	(9.4)	53.8	(21.5)
	High		14.1	(3.1)	12.0	(4.6)	28.3	(8.9)	19.3	(12.9)	73.7	(29.5)
High	Low		7.3	(1.6)	6.3	(2.4)	14.7	(4.6)	10.0	(6.7)	38.3	(15.3)
	High		10.1	(2.2)	8.6	(3.3)	20.2	(6.4)	13.8	(9.2)	52.7	(21.1)
Low	Low	Intermediate	13.2	(2.9)	11.2	(4.3)	26.4	(8.3)	18.0	(12.0)	68.8	(27.5)
	High		18.0	(4.0)	15.4	(5.8)	36.2	(11.4)	24.7	(16.5)	94.3	(37.7)
High	Low		9.4	(2.1)	8.0	(3.0)	18.9	(6.0)	12.8	(8.5)	49.1	(19.6)
	High		12.9	(2.8)	11.0	(4.2)	25.8	(8.2)	17.6	(11.7)	67.3	(26.9)
Low	Low	High	15.9	(3.5)	13.6	(5.2)	32.0	(10.1)	21.8	(14.5)	83.3	(33.3)
	High		21.8	(4.8)	18.6	(7.1)	43.8	(13.8)	29.9	(19.9)	114.1	(45.6)
High	Low		11.4	(2.5)	9.7	(3.7)	22.8	(7.2)	15.6	(10.4)	59.5	(23.8)
	High		15.6	(3.4)	13.3	(5.1)	31.3	(9.9)	21.3	(14.2)	81.5	(32.6)

[a]Estimated values for electricity demand (on a primary energy equivalent basis) have been placed in parentheses.

in the years 2000 and 2050, respectively, and therefore represent an upper bound for total electrical demand. If alternative energy forms are relied on to a greater extent for transportation, total demand for electricity will be correspondingly reduced.

SUMMARY OF UCS
DEMAND PROJECTIONS

Projected annual energy requirements for the year 2000 are in the range of 61-115 quads and for the year 2050, 38-114 quads. Even our upper case estimates—based on assumptions of low efficiency, high population growth, and high standards of living—fall well below estimates arrived at by historic extrapolations that reach values on the order of 190 quads per year by the turn of the century. Considering only the high efficiency cases, which we believe to be technically conservative, the range of projected energy demand is reduced to about 61-101 quads for the year 2000 and 38-81 quads for the year 2050. These estimates are considerably lower than those prepared by the federal government and energy industries, but are consistent with values obtained by recent analyses that have paid serious attention to energy efficiency considerations (CEQ 1979; CONAES 1978).

Even assuming a major role for electricity in transportation, our upper limit estimates for electrical energy requirements—attaining maximum levels of about 40 and 45 quads in 2000 and 2050, respectively—are well below previous government and utility forecasts based on a 7 percent annual growth that predicted a total demand of nearly 100 quads (primary energy equivalent) by the year 2000.

The salient points of this analysis are depicted in Figure 2-3. For illustrative purposes, four separate cases have been selected for examination, all of which assume the higher population growth rate. Curve A projects what future energy demands might be without improvements in energy efficiency, allowing for a 55 percent increase in per capita energy use. Curve B allows for the same increase in per capita energy use, but incorporates significant efficiency improvements. Paths A and B allow for identical levels of effective energy use and economic well-being. However, Curve A leads to a total annual energy use rate of 160 quads in 2050, whereas curve B leads to a level of about 80 quads or only one-half that rate. This comparison also demonstrates the fallacy of the frequently made assertion that energy conservation is a "one shot fix," implying that after the initial savings are attained, no additional benefits are derived. This is surely untrue. The development of new energy technologies and the phaseout of wasteful end uses can continue for many decades, and

Figure 2-3.   Future Energy Requirements.[a]

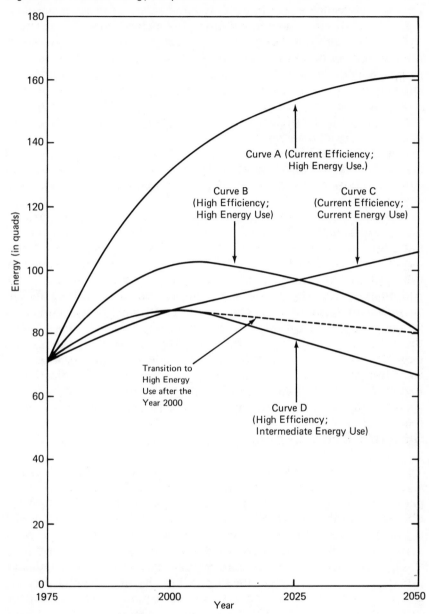

[a]All scenarios assume a high population growth rate.

the benefits, as we have seen, are cumulative. The net energy difference between the two paths illustrated is indeed massive, amounting in total to more than 3,000 quads through the year 2050. This is the energy equivalent of more than forty years of energy use at present U.S. rates and exceeds the total amount of energy estimated to be available in domestic resources of oil or uranium (assuming the latter is "burned" in conventional light water reactors). Thus, it is clear that energy efficiency constitutes a highly significant energy "resource"; minor efficiency improvements can obviate the need for major expansion in energy supplies.

The difference between curves B and C are also striking. Curve B allows for a much higher level of effective per capita energy use than curve C yet, owing to efficiency improvements, uses considerably less energy in 2050.

Overall energy demands are reduced even further for curve D, which incorporates the same efficiency improvements as curve B, but assumes less ambitious increases in per capita energy use. While the high standard of living levels may not be practically attained by the turn of the century, this goal should be achievable by 2050. Although energy supply does not appear to be a limiting factor, environmental, resource, and/or capital constraints to the implementation of these changes need to be more fully assessed. This suggests the attractive possibility of following the intermediate energy use path of curve D through the year 2000 and then making a smooth transition to the higher energy use path of curve B (as indicated on Figure 2–3 with a dashed line). In this way, the sharp early peak in energy demand, inherent to path B, would be circumvented.

GENERAL CONCLUSIONS

This admittedly simple analysis of future energy demand does not attempt to yield precise predictions, but rather to chart out possible paths for the nation to travel. Many refinements in procedure are obviously possible. Nonetheless, a conclusion that powerfully emerges is that the United States can provide a high level of economic prosperity for all its citizens without the wastefully high levels of gross energy use characteristic of conventional government and industry projections. By increasing energy productivity and thereby deriving greater benefit from the energy we use, economic growth can be sustained with little or no growth in overall energy use. Owing to the vast potential for purely technical improvements in the efficiency of energy utilization, a reduction in energy growth can occur without any attendant reduction in the quantity or quality of energy services provided and without the need for personal sacrifices or cutbacks in living standards. Saving energy through improved end use efficiency is not only cheaper than extending energy supplies, but is also much less taxing on the environment. Moreover, by slowing the rate at which our premium fuels are exhausted, we can "buy" additional time so helpful for a thoughtfully planned shift to a reliance on abundant, renewable energy resources. In light of the compelling advantages, improved energy efficiency clearly deserves foremost consideration in both near-term and long-range national energy planning.

REFERENCES

American Institute of Architects (AIA). 1975. "A Nation of Energy Efficient Buildings by 1990." New York.

American Physical Society (APS). 1975. "Efficient Use of Energy: A Physics Perspective." New York, January.

Atomic Energy Commission (AEC). 1974. *Liquid Metal Fast Breeder Reactor Program*. Final Environmental Impact Statement. Vol. 4. Washington, D.C., December.

Ayres, R., and M.N. Kramer. 1976. "An Assessment of Methodologies for Estimating National Energy Efficiency." New York: American Society of Mechanical Engineers, Paper #76-WA/TS-4.

Bliss, R.W. 1976. "Why Not Just Build the House Right in the First Place?" *Bulletin of the Atomic Scientists*, March, pp. 32-40.

Bureau of the Census. 1975a. "Population Estimates and Projections." *Current Population Reports*. Series p-25, no. 601. February.

———. 1976. "Statistical Abstract of the United States." Washington, D.C.: U.S. Department of Commerce, July.

Committee on Nuclear and Alternative Energy Systems (CONAES). Demand and Conservation Panel. 1978. "U.S. Energy Demand: Some Low Energy Futures." *Science* 200 (April 14):142-52.

Cook, E. 1976. *Man, Energy, Society*. San Francisco: W.H. Freeman and Company.

Council on Environmental Quality (CEQ). 1978. "Environmental Quality— Ninth Annual Report." Washington, D.C.

———. 1979. "The Good News About Energy." Washington, D.C.

Department of Commerce. 1971. "The U.S. Energy Problem." Prepared by Intertechnology Corporation. Vol. 1. Washington, D.C.

Department of Energy (DOE). 1978. Washington, D.C.: Energy Information Administration, DOE/EIA-0036-2.

———. 1979a. "Monthly Energy Review." Washington, D.C.: Energy Information Administration, DOE/EIA-0035-7(79). July.

———. 1979b. "Annual Report to Congress—1978." Vol. 2. Washington, D.C.: Energy Information Administration, DOE/EIA-0173/2.

Department of the Interior. 1976. "Annual U.S. Energy Use Drops Again." News Release. Washington, D.C., April 5.

Department of Transportation. 1975. "Summary of National Transportation Statistics." Report no. DOT-TSC-OST-75-18. Washington, D.C., June.

Dow Chemical Company. 1975. "Energy Industrial Center Study." Report no. PB-243-824. Midland, Mich., June.

Dupree, N.G., and J.A. West. 1972. "United States Energy Through the Year 2000." Washington, D.C.: Department of the Interior, December.

Ehrlich, P.R.; A.H. Ehrlich; and J.P. Holdren. 1977. *Ecoscience: Population, Resources, Environment.* San Francisco: W.H. Freeman and Company.

Eketorp, S. 1978. "Decisive Factors for Planning Future Steel Plants." Paper presented at the Third International Iron and Steel Congress, Chicago, Illinois, April.

Energy Policy Project of the Ford Foundation (EPP). 1974. *A Time to Choose.* Cambridge, Mass.: Ballinger Publishing Company.

———. 1974b. *Energy Consumption in Manufacturing.* Cambridge, Mass.: Ballinger Publishing Company.

Energy Research and Development Administration (ERDA). 1975. "A National Plan for Energy Research, Development, and Demonstration: Creating Energy Choices for the Future." ERDA-48. Vol. 1. Washington, D.C.

Exxon Company. 1975. "Energy Outlook: 1975-1990." Houston.

Federal Energy Administration (FEA). 1974. "Project Independence Report." Washington, D.C.

Goen, R.L., and R.K. White. 1975. "Comparison of Energy Consumption Between West Germany and the United States." Menlo Park, Calif.: Stanford Research Institute, Report no. PB-245 652, June.

Halloran, R. 1979. "Optimism on Fuel Energy." *New York Times*, February 21, p. D1.

Hayes, D. 1977. *Rays of Hope—The Transition to a Post-Petroleum World.* New York: W.W. Norton and Company.

Herendeen, R.A. 1974. "Affluence and Energy Demand." *Mechanical Engineering*, October.

Higgs, F.S., ed. 1978. "Special Princeton Issue." *Energy and Buildings* 1, no. 3 (April).

Hirst, E. 1973. "Energy Intensiveness of Passenger and Freight Transport Modes: 1950-1970." Oak Ridge, Tenn.: Oak Ridge National Laboratory, ORNL-NSF-EP-44, April.

Institute of Energy Analysis (IEA). 1976. "Economic and Environmental Impacts of a Nuclear Moratorium: 1985-2010." Oak Ridge, Tenn.: Oak Ridge Associated Universities, ORAU/IEA 76-4, September.

Karkheck, J., et al. 1977. "Prospects for District Heating in the U.S." *Science* 195 (March 11):948-55.

Keeny, S.M., et al. 1977. *Nuclear Power Issues and Choices.* Report of the Nuclear Energy Policy Study Group. Sponsored by the Ford Foundation, administered by the MITRE Corporation. Cambridge, Mass.: Ballinger Publishing Company.

Metz, M.D., and A.L Hammond. 1978. *Solar Energy in America*. Washington, D.C.: American Association for the Advancement of Science.

McElheny, V.K. 1978. "The Energy Efficiency of U.S. Industries." *New York Times*, March 26.

Nader, R. 1976. Statement Before the Senate Commerce Committee. February 26.

National Academy of Sciences (NAS). 1976. "Criteria For Energy Storage R&D." Washington, D.C.

New York Times. 1979. "U.S. Shift From Oil Projected." February 26.

Nydick, S., et al. 1976. "A Study of the Inplant Electric Generation in Chemical, Petroleum Refining, and Paper and Pulp Industries." Prepared for the Federal Energy Administration by the Thermo Electron Corp., Waltham, Massachusetts, May.

Rice, R.A. 1974. "Toward More Transportation With Less Energy." *Technology Review*. February.

Ross, M.H., and R.H. Williams. 1975. "Assessing the Potential for Fuel Conservation." Stony Brook, N.Y.: The Institute for Public Policy Alternatives, State University of New York, July.

———. 1976. "Energy Efficiency: Our Most Underrated Energy Resource." *Bulletin of Atomic Scientists*, November.

———. 1977. "Energy and Economic Growth." In *Energy*, vol. 2. Paper no. 2. A Study Prepared for the Use of the Joint Economics Committee, Congress of the United States. Washington, D.C.: Government Printing Office, August 31.

Schipper, L. 1975. "Holidays, Gifts, and the Energy Crisis." Berkeley, Calif.: Energy and Resources Group, University of California—Berkeley, ERG 75-11.

Schipper, L., and A.J. Lichtenberg. 1976. "Efficient Energy and Well-Being: The Swedish Example." *Science* 194 (December 3).

Steel '78 (published by the American Iron and Steel Institute). 1978. "Facing the Energy Crunch Through Conservation."

Twain, M. 1961. *Life on the Mississippi*. New York: The New American Library of World Literature, Inc.

Washington Post. 1977. "Energy Efficient Germany." June 20.

White House. 1979. "Fact Sheet on the President's Program." Washington, D.C.: Office of the White House Press Secretary, April 5.

Widmer, T.F., and E.P. Gyftopolous. 1977. "Energy Conservation and a Healthy Economy." *Technology Review*, June, pp. 31–40.

Williams, R.H. 1978. "Industrial Cogeneration." *Annual Review of Energy* (Annual Reviews, Inc.) 3:313–56.

Chapter 3

Nonrenewable Energy Resources

In this chapter, nonrenewable energy resources—fossil fuel, nuclear, and geothermal—are evaluated. The supplies of these depletable resources are limited and essentially non-replenishable. For example, the fossil fuels now available for use—coal, oil, and natural gas—were accumulated over hundreds of millions of years, storing solar energy initially captured by plants ages ago. These fuels are formed at rates that are insignificant in comparison to the present rate of their extraction and utilization. Indeed, one of the most grave components of the energy problems facing the United States is the imminent depletion of its domestic oil and natural gas and, followed by at most a few decades, the depletion of the world supplies. Nuclear fission and fusion exploit the energy locked up in atomic nuclei, an energy source that, although potentially very large, is nonetheless finite. Geothermal energy, the energy stored within the earth, represents a theoretically vast resource that is somewhat more difficult to classify. Nonetheless, it must be regarded as non-renewable. The energy that is most convenient for utilization—that contained in pockets of hot water and steam trapped near the earth's surface—is quite limited in supply and would only slowly be re-generated from the larger reservoir of thermal energy stored in the earth's core. The heat that is continually produced by the nuclear decay of radioactive elements within the earth's crust is too diffuse and is generated at too slow a rate for practical applications.

One intrinsic feature of nonrenewable resources is the fact that the more they are developed and the closer they approach exhaustion, the harder and costlier it becomes to locate and exploit new sources. Thus, a continued dependence on a depletable resource inevitably means a steady increase in its production costs. In reality, a nonrenewable resource would never be completely exhausted. Well before total depletion occurred, further exploitation would become economically, and perhaps technically, unfeasible.

The United States is almost exclusively dependent on nonrenewable energy sources at the present time. As shown in Table 3-1, nonrenewable fuels provided roughly 96 percent of total energy use in 1978, or, more precisely, 94 percent when the 1.8 quads of wood consumed by residential and industrial consumers is considered (DOE 1978a). Petroleum and natural gas provide nearly three-fourths of the total U.S. energy supply and together with coal account for 90 percent of domestic energy use. Nuclear power generation contributed slightly less than 4 percent. The remaining 6 percent of energy use in 1978 was provided by renewable sources—hydroelectric power generation and the burning of wood, plus an extremely modest contribution from geothermal power.

An estimate of the amount of energy potentially available from the various possible resources and an evaluation of the relative merits and disadvantages of each is an indispensable guide to future energy

Table 3-1.   Energy Use by Source and Consuming Sector—1978 (in quads).

	Residential and Commercial	Industrial	Transportation	Electric Utilities	Total	Percent of Total
Coal	0.3	3.4	—	10.4	14.1	18.0
Petroleum	6.4	7.6	20.0	3.9	37.9	48.5
Natural Gas	7.7	8.3	0.5	3.3	19.8	25.3
Hydroelectric	—	—	—	3.1	3.1	4.0
Nuclear	—	—	—	3.0	3.0	3.8
Other	—	0.2[a]	—	0.1[b]	0.3[c]	0.4
Total (with electricity disaggregated)	14.4	19.5	20.5	23.8	78.2	100.0
Total (with electricity)	28.6	29.0	20.6	—	78.2	100.0

[a]Includes coke imports by steel industries and on-site hydroelectric power generation.
[b]From geothermal, wood, and waste sources.
[c]Does not include an estimated 1.8 quads of wood consumed by residential and industrial users.
Source: DOE (1979).

planning. By clearly identifying resource, technology, economic, and environmental limitations, it will be possible to develop a realistic national energy strategy that can smoothly and successfully carry us through a transition to a sustainable energy future. Estimates of the energy potentially recoverable from nonrenewable sources are presented in Table 3-2, and illustrated in Figure 3-1, followed by a brief summary of our assessment of these sources. More detailed reviews of the nonrenewable resources of significance for the United States are presented in subsequent sections.

Coal, in abundant supply in the United States, will be a necessary and important transitional fuel. However, the numerous environmental problems involved in the mining, transportation, and burning of coal will impose stringent standards on its utilization. New technologies designed to produce synthetic oil and gas from coal appear likely to augment, rather than mitigate, problems relating to coal utilization owing to their high capital costs, inherent conversion inefficiencies, and the consumption of prodigious amounts of water necessary for production of these fuels. In the long term, the use of coal may have to be curtailed well before supplies are exhausted in order to prevent significant climatic modification from cumulative additions of carbon dioxide from its combustion to the atmosphere.

Oil and gas, together providing nearly 75 percent of the U.S. energy supply, are fuels of critical importance to the nation's economy. Unfortunately, they are nonreplenishable and in limited supply. Although, the quantities of U.S. oil and gas resources that can be economically recovered are not known with great precision, it appears that with prudent and efficient utilization, these resources should be capable of supporting continued use until sometime in the next century. Owing to their present importance and relative attractiveness from an environmental standpoint, oil and gas will necessarily play a crucial role in the transition to a sustainable energy future. Therefore, more accurate information regarding the amounts and production costs of remaining supplies is desperately needed in order to formulate an intelligent transitional strategy.

Oil shale resources in the United States are literally enormous and could conceivably extend supplies of domestic petroleum for many decades at a minimum. However, oil shale production on an extensive basis would lead to substantial environmental problems including the severe disruption of previously undisturbed land, the generation of huge volumes of waste by-products, and the consumption and contamination of vast quantities of water in arid regions. Furthermore, the development of a large-scale oil shale industry will require significant investments of capital. Oil shale

Table 3-2. Nonrenewable Energy Resource Estimates.

Nonrenewable Resource	*Total Nonrenewable Energy (in quads)*
Fossil: Coal	6,000–21,000[a]
Liquid Petroleum	800–1,200[b]
Natural Gas	700–1,200[c]
Oil Shale	500–1,200[d]
Nuclear Fission: Light Water Reactors	700–2,400[e]
Breeder Reactors	40,000–1,000,000[f]
Nuclear Fusion	200,000–1,000,000,000[g]
Geothermal	300–1,000[h]

[a]*Coal*

Low Estimate—6,000 quads. Based on Bureau of Mines estimate of about 430 billion tons of coal in minable deposits, 250 billion tons of which are expected to be recoverable (Hayes 1979).

High Estimate—21,000 quads. This value, which appears to represent a maximum upper limit, is based on an estimate by ERDA (1976a) of 850 billion tons of coal, including demonstrated reserves plus additional resources, all of which may not be recoverable. A more credible upper limit may be 13,000 quads, based on an ERDA (1976a) estimate of 530 billion tons of recoverable coal resources.

[b]*Liquid Petroleum*

Low Estimate—800 quads. Based on an estimate by Hayes (1979) of 37 billion barrels of proven reserves of petroleum liquids (crude oil and natural gas liquids) plus an additional 100 billion barrels of undiscovered resources. This value is also consistent with estimates of onshore and offshore resources made by Hubbert (1974) and the Geological Survey (USGS 1975a), assuming a historical recovery factor for oil of 32 percent.

High Estimate—1,200 quads. Assuming that about 200, rather than 140, billion barrels could be made available from proven reserves and undiscovered petroleum resources with an enhanced recovery factor of 45 percent.

[c]*Natural Gas*

Low Estimate—700 quads. Based on an estimate by Hayes (1979) of 200 trillion cubic feet (tcf) of proven reserves and 500 tcf of undiscovered gas resources. This figure is also consistent with an estimate made by Hubbert (1974) for onshore and offshore gas resources, assuming a historical recovery factor of 80 percent.

High Estimate—1,200 quads. Based on an estimate by Mankin (1979) of gas resources in onshore and offshore deposits. Included in this estimate are natural gas resources assumed to be recoverable from "unconventional" sources—deep deposits, shale, and tight sands. A more conservative upper limit of the total onshore and offshore gas resource (excluding regions off the Atlantic and Alaska coast) is provided by a USGS (1975a) estimate of 900 quads.

[d]*Oil Shale*

Low Estimate—500 quads. Assumes that 80 to 90 billion barrels of shale oil are recoverable from high grade deposits in the Green River Basin (Bureau of Mines 1975; Hayes 1979; National Petroleum Council 1972). The Green River formations, underlying parts of Colorado, Wyoming, and Utah, hold the richest identified deposits in the United States; the total amount of shale oil (only a small fraction of which can be recovered) has been estimated to be about 2 trillion barrels (Dinneen and Cook 1972; Hayes 1979).

High Estimate—1,200 quads. Based on the extraction of 200 billion barrels of high grade shales from the Green River Basin (ERDA 1976a)—approximately equal to 10 percent of the total estimated shale resource in that formation.

[e]*Nuclear Fission—Light Water Reactor (LWR)*

Low Estimate—700 quads. Based on an estimate by ERDA (1976b) of 1.8 million tons of high grade uranium oxide resources consisting of about 800,000 tons of proven reserves and an additional 1 million tons of probable resources. This is enough uranium to fuel approximately 400 light water reactors operating at an average capacity factor of 65 percent over an average lifetime of thirty years, assuming that uranium requirements per reactor year of about 143 tons per year, typical of currently operating LWRs without plutonium recycle (Feiveson, von Hippel, and Williams 1979).

High Estimate—2,400 quads. Based on the Department of Energy (DOE 1978) estimate of 4.4 million tons of recoverable uranium oxide resources and assuming that advanced LWRs operating with uranium-plutonium recycle would derive 40 percent more energy per unit of fuel consumed than present-day reactors operating without fuel recycling.

[f]*Nuclear Fission—Liquid Metal Fast Breeder Reactor (LMFBR)*

Low Estimate—40,000 quads. Based on the ERDA (1976b) estimate of 1.8 million tons of high grade uranium oxide reserves and probable resources and assuming that LMFBRs utilize uranium approximately sixty times more efficiently than do LWRs.

High Estimate—1,000,000 quads. Assumes the utilization of low grade uranium oxide resources (including the Chatanooga Shales and Conway Granits), amounting to more than 40 million tons, in addition to high grade resources. However, excessive environmental costs may preclude the extraction of low grade uranium oxide resources. If only high grade resources are utilized, assuming the DOE (1978) estimate of 4.4 million tons as an upper limit, the total energy potential would be reduced to about 100,000 quads.

[g]*Nuclear Fusion*

Low Estimate—200,000 quads. This estimate is for deuterium-tritium (D-T) fusion, based on reasonably assured U.S. supplies of lithium, which is the limiting resource for this reaction (Holdren 1978). If the more technically difficult proton-boron[11] reaction could be achieved, U.S. reserves of boron would be sufficient to provide about 2 million quads from nuclear fusion (Holdren 1978).

High Estimate—1 billion quads. For the proton-boron[11] fusion reaction, assuming that the United States uses about 0.5 percent of the boron resources estimated to be recoverable from the world's oceans (Holdren 1978). The potential energy yield of deuterium-deuterium (D-D) fusion is even greater, but radioactivity problems stemming from high neutron fluxes and the possibility for direct energy conversion make the proton-boron reaction much more attractive.

[h]*Geothermal Energy*

Low Estimate—300 quads. Includes all identified reserves of intermediate and high temperature liquids and vapors accessible through hydrothermal convection and all identified geopressurized liquid and methane reserves that can be extracted with minimal subsidence (USGS 1975b).

High Estimate—1,000 quads. Includes all identified reserves of hydrothermal liquids and vapors plus one-half of the estimated undiscovered resources in these categories (USGS 1975b). It is further assumed that the undiscovered geopressurized gas resource that can be extracted in an environmentally acceptable manner is equal in magnitude to the identified resource.

Although supposedly vast quantitites of geopressurized methane underlie the Gulf Coast region, recovery could be impeded by a formidable array of technical, economic, and environmental obstacles. Hot dry rock and magma geothermal energy sources also have not been included in our estimates. According to one estimate (Cummings et al. 1979), extraction of 0.1 percent of the geothermal energy contained in crustal rock underlying the United States within six miles (10 km) of the surface could provide more than 10,000 quads—equivalent to a reasonable estimate of coal resources. Although these resources are potentially very large, the appropriate extraction technology does not yet exist, and we have thus made no attempt to extrapolate beyond known technology.

Figure 3-1. Nonrenewable Energy Resource Estimates. The height of the bars reflects the uncertainty in current estimates. A discussion is provided in Table 3-2 and associated notes.

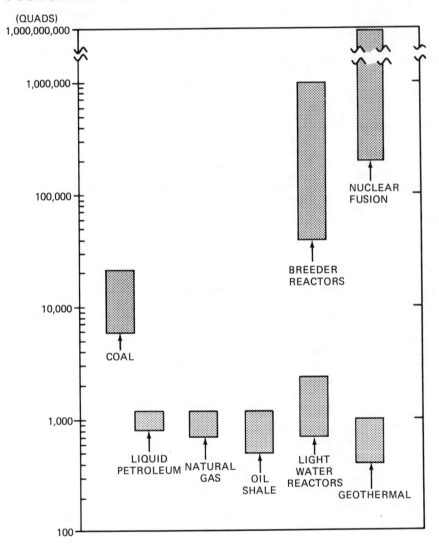

production is not economically feasible at present and appears to be prohibitively expensive at least for the foreseeable future. Therefore, it is unlikely that oil shale will make an appreciable contribution to the nation's energy supply in this century. It will become an important energy source only at the expense of neglecting more benign alternatives.

Nuclear power, once regarded as the "ultimate" energy solution, is presently besieged with a bewildering array of problems that considerably dampen its future prospects. Growing doubts over nuclear safety have been recently reinforced by the March-April 1979 accident at the Three Mile Island plant in Pennsylvania. Radioactive waste disposal is proving to be much more difficult than was indicated by earlier, optimistic pronouncements. Finally, nuclear plant construction and fuel cycle costs have escalated sharply in recent years, seriously threatening any economic advantage this power source might once have held. A combination of these and other factors has contributed to a virtual halt in the ordering of new nuclear power plants, with cancellations vastly exceeding new orders over recent years.

In addition to these problems, the ultimate potential of present U.S. nuclear technology, based on light water reactors, is inherently constrained by limited domestic supplies of uranium fuels. In fact, U.S. resources of the high grade uranium ores "burned" in present-day light water fission reactors may only be capable of providing as much energy as domestic supplies of oil, themselves in short supply. The breeder reactor, which would produce more usable fuel than it consumes, could dramatically stretch the amount of energy that could be derived from nuclear fission. Unfortunately, the breeder inherits most, if not all, of the problems inherent to light water reactors and in some cases heightens, rather than alleviates, their severity. Breeder reactors are almost certain to be more expensive than present reactor technology. In addition to inheriting problems of reactor safety and radioactive waste disposal, a breeder reactor would require fuel reprocessing—a "messy" business that has so far proved commercially unsuccessful. Owing to the vast quantities of highly enriched fissile material flowing through a breeder reactor system, nuclear proliferation threats and the risks of nuclear terrorism would be greatly increased in a large-scale breeder economy. In light of these difficulties, we believe a major commitment to breeder development would be unwise at the present time.

Nuclear fusion offers the possibility of providing virtually unlimited amounts of energy. However, efforts to harness fusion power to date have not succeeded in producing a workable reactor. While there is no reason to believe the achievement of controlled nuclear fusion to be intrinsically impossible, neither can its technological or economic feasibility be assured. Therefore, the nuclear fusion option can neither be counted on nor responsibly dismissed at the present time.

Geothermal energy, encompassing the vast reservoirs of energy stored within the earth, curiously offers little energy potential for this century. The amount of energy contained in accessible geothermal deposits that are recoverable with present technology is, in fact, quite limited. The extraction of much greater amounts of energy from more remote geothermal sources awaits the development of new technology. The extent to which this energy could ever be economically exploited within environmentally acceptable limits is not yet known.

COAL

Early in this century, coal served as the major source of energy for the United States. Owing to the availability of energy in the more convenient forms of oil and natural gas, coal production in this country has actually declined since 1920. With the depletion of domestic supplies of oil and gas now imminent, hopes have once again turned to exploiting the vastly more abundant reservoirs of coal, representing almost 90 percent of all proven domestic fossil fuel reserves.

Expansion of coal production in the future, however, will be impeded by inherent difficulties involved in its mining, transportation, and utilization. In the long term, the burning of coal, as well as other fossil fuels, may have to be curtailed to avoid significant global climatic modifications.

In 1978, coal provided about 14 quads of primary energy, accounting for nearly 18 percent of total U.S. energy use. Coal is presently a major source of electrical generation, providing for nearly 45 percent of the electricity produced in the United States in 1978. Of all the coal currently consumed, nearly two-thirds is devoted to electricity generation, with the bulk of the remainder used for industrial process heat and steam applications.

The potential coal resources of the United States are, by any measure, enormous. The United States possesses about 31 percent of the world's known coal reserves. Proven reserves—those resources positively identified—amount to some 430 billion tons (10,750 quads),[1] of which at least 250 billion tons, or 6000 quads, are re-coverable with currently available technology (Hayes 1979). Potential coal resources may be as high as 21,000 quads, but all of this may not be recoverable (ERDA 1976a). At current domestic rates of coal consumption, currently recoverable reserves would last at least 400 years. If coal were to provide all our energy needs, these reserves would last for more than seventy-five years at current rates of total U.S. energy consumption.

The production of coal, in light of its abundance, is projected to increase throughout the remainder of this century. Increased use of coal as an energy source will require the implementation of new com-bustion techniques, because traditional methods of coal burning will be unable to meet future air quality standards. Flue gas desulfuriza-tion, or "scrubbing," is being widely implemented to cut down on sulfur oxide emissions from coal-fired electric generating plants. Fluidized bed combustion, in which pulverized coal is mixed with limestone, suspended in air, and burned, is also able to reduce sulfur emissions by as much as 90 percent (Hammond 1976). However, neither technology eliminates the emission of small ash particulates, which constitute a serious health hazard.

Coal can also be converted into liquid or gaseous fuels. The syn-thetic petroleum produced by coal liquefaction methods is of suffi-cient quality to be refined into gasoline and diesel fuels. Gasification of coal can produce either a "low Btu" gas, suitable for use by plants located close to the production site, or higher-quality methane, which can be transported by long distance pipeline.

These fuels may be burned with far less environmental and health impacts than the coal they were initially made from. However, the environmental impacts incurred in their production are far from trivial. Furthermore, synthetic fuel production suffers from inherent inefficiencies: coal mining would have to be increased by about 50 percent to provide the energy equivalent of current coal consumption in the form of synthetic oil and gas (Wilson 1977). The technology for coal liquefaction and gasification has not yet been demonstrated on a large scale in the United States. The anticipated costs of fuels produced by these new technologies constitute the major barrier to development of either on a commercial scale.

In fact, although coal-fired electric power plants presently have lower capital costs than nuclear plants of equal size, the economics of many presently new coal-based technologies are much less favor-able. For example, liquefaction of coal yields fuels costing more than

---

1. Forty million tons of bituminous coal is equivalent to 1 quad.

Map 3-1:   Distribution of Coal Resources in the United States.

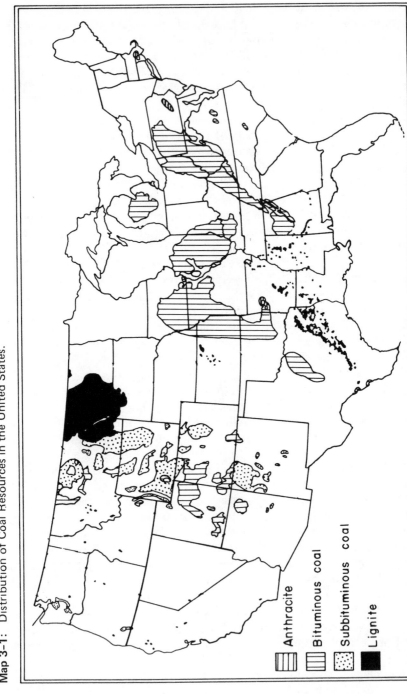

Anthracite

Bituminous coal

Subbituminous coal

Lignite

Source: Bureau of Land Management. "Draft Environmental Impact Statement: Proposed Federal Coal Leasing Program." Washington, D.C.: Government Printing Office, 1974. p. I-47.

Photograph 3-1: Coal Gasification. The SYNTHANE Pilot Plant, near Pittsburgh, converts 72 tons of coal into 1.2 million cubic feet of gas each day, part of which is methanated to pipeline quality substitute natural gas. The plant has been in operation since 1976.

Source: Department of Energy.

$30 per barrel (Hayes 1979), substantially more expensive than OPEC oil, reflecting not so much the high costs of the original coal resource, but rather the great capital costs of the conversion technologies. Eventually, the rapidly increasing price of natural petroleum is likely to exceed the cost of synthetic fuels, but until then, widespread application of liquefaction and gasification is not likely to take place.

Another set of obstacles to increased coal consumption is posed by its adverse safety and environmental effects. Underground coal mining is still one of the most hazardous of industrial occupations in spite of the fact that existing technologies have been implemented at a modest cost in a small number of mines so as to reduce the accident rate. Strict enforcement of federal dust standards could virtually eliminate black lung disease, responsible for the incapacitation of large numbers of coal miners. Coal mining also has serious effects on land quality. Strip mining denudes large tracts of land, and underground mining can result in severe land subsidence. Although some rehabilitation of stripped lands is possible, land quality is almost never fully restored to its original condition. Coal liquefaction and gasification consume and contaminate large quantities of water, a problem of particular significance in western states with limited water resources where many of the coal reserves are located.

More problematic environmental effects arise from emissions of gaseous and particulate matter and carbon dioxide resulting from coal combustion. These problems, although most significant for coal, are inherent to all other fossil fuels as well. Emissions of sulfur dioxide from coal- and oil-fired plants and nitrogen oxides from the combustion of all fossil fuels increase the acidity of precipitation, leading to the phenomenon of "acid rain." This problem is particularly acute in the northeastern United States, where fish populations have decreased in many lakes and reductions in forest and agricultural productivity are suspected (CEQ 1978). Dust and other particulates produced by the burning of coal absorb and scatter radiation, reducing the amount of sunlight reaching the earth's surface. Most troublesome is the problem of carbon dioxide—an inevitable by-product of fossil fuel combustion. Carbon dioxide in the atmosphere has increased by about 15 percent over the past one hundred years, and since 1950, the worldwide use of fossil fuels has increased at an annual rate of 4.3 percent. The carbon dioxide content of the atmosphere is expected to double within the next fifty to one hundred years (MacDonald 1979). Carbon dioxide in the atmosphere is transparent to visible light, but absorbs infrared radiation or heat re-radiated by the earth into space, contributing to a net warming of the earth's surface. This is called the "greenhouse effect." Models suggest that as a result of the doubling of atmospheric carbon dioxide content, average surface temperatures will rise some 3.5 to 5.5°F, with an amplified effect at higher latitudes (MacDonald 1979). These increased temperatures are likely to have severe effects upon global climates, affecting agricultural production, the availability of water, and the sea level at coastal cities.

A problem common to all nonrenewable energy sources is thermal pollution. The release of large amounts of waste heat into the atmosphere could also result in a long-term global warming trend. If world energy consumption were to increase at an annual rate of 4 percent over the next 125 years, the earth's surface temperature would rise by several degrees Fahrenheit (Kellogg 1975). The magnitude of these global effects on climate must be more accurately assessed before large future commitments to any nonrenewable energy source— coal, nuclear fission or fusion, or geothermal—are made.

Although coal will continue to be a necessary energy source, at least for the next several decades, its use will have to be subject to stringent environmental standards. Over the longer term, a large-scale expansion in coal consumption may prove to be unacceptable due to the possibility of irreversible climatic modification combined with other adverse environmental impacts.

PETROLEUM

Liquid petroleum is literally the lifeblood of the modern industrial state, and its availability has helped make possible the dramatic advances in industrialization occurring in this century, particularly within the past several decades. Petroleum's role in transportation has been especially critical: over 95 percent of the energy presently used for transportation in the United States is derived from petroleum. The impending decline in domestic and world oil production will force major shifts in future transportation systems as well as in other enery use sectors. Another important problem is the growing U.S. dependence on imported foreign oil. Mounting instabilities in international supplies and steadily increasing prices make a reduced dependence on oil imports an economic necessity.

In 1978, the United States consumed roughly 6.5 billion barrels, or about 38 quads, of petroleum, which provided 48 percent of primary domestic energy use. Petroleum imports made up 49 percent of domestic requirements. Of total petroleum consumption, only about 10 percent is used in the generation of electricity, with the remaining 90 percent being devoted to transportation, thermal applications, and petrochemical feedstocks. Overall, about 53 percent of domestic petroleum consumption goes to the transportation sector, 17 percent to non-electrical residential and commercial uses, and 20 percent to industrial applications.

In contrast to coal, petroleum reserves are quite limited. Measured domestic reserves of crude oil and natural gas liquids totaled 35 billion barrels, or 203 quads[2] (*New York Times* 1979), down from a peak of 47 billion barrels (273 quads) in 1970 (Hayes 1979). Depending upon assumed recovery factors, undiscovered domestic resources are estimated to be roughly between 100 billion barrels (Hubbert 1974; Hayes 1979) and 200 billion barrels (USGS 1975a). Together, remaining liquid petroleum reserves and resources lie in the range of 800 to 1200 quads, equivalent to a twenty to thirty-five year supply at current rates of petroleum consumption or a forty to seventy year supply if imports are assumed to continue to comprise about 50 percent of domestic requirements. However, it is far from certain that oil can be produced at a rate to keep pace with continued demand.

Foreign oil resources are quite large, but are subject to a variety of geopolitical and technical factors that could, in the future, restrict their exploitation. At the end of 1976, recoverable world reserves of crude oil stood at 567 billion barrels (Hayes 1979). Prospective and undiscovered global resources are believed to exceed 2,000 billion barrels (UN 1974), with as much as 560 billion barrels in the Middle East alone. More recently, the potential petroleum resources of Mexico have been estimated to range anywhere from 200 to 700 billion barrels (Metz 1978). Despite these large numbers, there are considerable obstacles to a major and rapid expansion of worldwide oil production. In addition, competition from other nations for world oil supplies will tend to place limits on the availability of oil to the United States and on the length of time world supplies will last. At current global consumption rates of some 20 billion barrels per year, supplies would last no longer than one hundred years, even if total worldwide resources, including those still undiscovered, are recovered—a prospect whose technical, economic, and political feasibility is by no means assured.

Contrary to a commonly held view, oil does not exist in vast underground pools that can simply be drained dry. It is generally found trapped and dispersed in porous rock formations, in a manner analagous to water droplets in a sponge, bounded on all sides by non-porous layers. Within the trap, three distinct layers usually occur—

2. One billion barrels of oil is roughly equivalent to 5.8 quads.

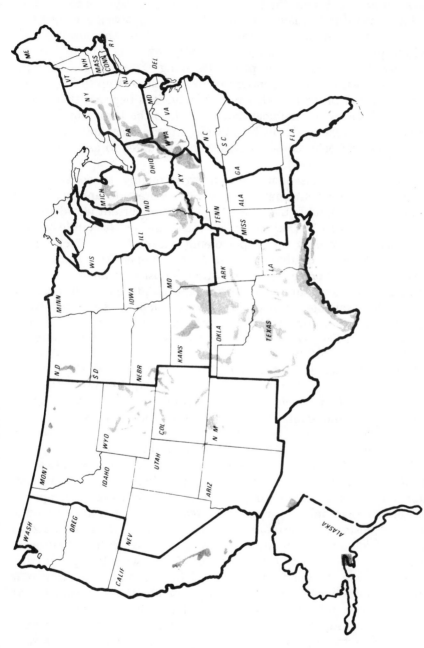

Map 3-2: Oil and Gas Fields of the United States.

Source: Crump, L.H. "Fuels and Energy Data: United States by States and Census Divisions, 1974." Bureau of Mines Information Circular 8739. Washington, D.C.: Government Printing Office, 1977. p. 11.

gas-saturated rock near the top, an oil-containing layer in the middle, and a brine-saturated layer at the bottom. When a well is drilled into the trap, pressure in the reservoir forces oil to the surface. This "primary recovery" technique generally extracts about twenty-five percent of the oil in place. Pumping water or gas into the trap—a process called "secondary recovery"—allows up to 32 percent of the original oil to be extracted. Finally, it is believed that the injection of steam or chemical detergents into the trap, which would facilitate the release of oil by lowering its viscosity and allowing it to flow more freely, could increase recovery rates to as much as 45 percent (Tiratsoo 1973).

The preceding discussion has referred mainly to conventional, "light"oil. Unconventional, "heavy" oil represents another major resource that accounts for about 10 percent of current worldwide oil production (Parisi 1979a). Heavy oil is a dense, viscous crude oil that is similar to ordinary liquid petroleum except for the fact that it must be heated or diluted before it can be pumped from the ground. Lying even lower on the spectrum of oil deposits are tar sands, which consist of grains of sand coated with semisolid, tarlike hydrocarbons. Potentially available supplies of heavy oil and tar sands are large: deposits exceeding one trillion barrels have been identified in both Venezuala and Canada, while known deposits in the United States are on the order of 200 billion barrels (Marshall 1979; Bureau of Mines 1979). It has been estimated that 15 billion barrels of this U.S. heavy oil could be practically recovered (Parisi 1979a). In contrast to light oil, heavy oil is a largely untapped resource—a fact directly attributable to the unfavorable economics of production that have prevailed to date. The economic outlook will become increasingly favorable, however, as the price of conventional oil continues to rise.

The era of "cheap" oil came to an abrupt end in 1973–1974 when the Organization of Petroleum Exporting Countries (OPEC) quadrupled the world price of oil. In a real sense, however, "cheap" oil was inexpensive only in that it reflected the average rather than the marginal costs of oil production. As each barrel of oil is extracted from a field, the remaining oil becomes more difficult and more expensive to produce. As established oil fields, with low production costs, become depleted, the costs of finding and extracting new supplies of oil inevitably increase. Traditionally, these increased costs have not been reflected in the price of petroleum, as the costs of old and new supplies have been lumped together into an average value. Furthermore, oil is a resource that, once consumed, is gone forever. The irreplaceable nature of the oil resource increases the difficulty of determining a rational pricing policy. While uncertainties abound over the future price of oil, it is clear that prices will continue to rise.

Environmental impacts may place some constraints on the future production, transportation, and combustion of petroleum, but these problems are, on balance, of a considerably lesser magnitude than those associated with coal. Petroleum extraction is a safer activity and causes less insult to land than coal mining. The utilization of tars sands, however, would constitute a notable exception to this general rule. Being a highly dilute source of petroleum, large-scale strip mining would be required to extract oil from the sands, unless underground, "in situ" recovery methods were employed. In any event, both tar sand and heavy oil production would consume and befoul large quantities of water; it has been estimated that about 5 barrels of fresh water would be consumed per barrel of oil produced at one tar sand project proposed for Alberta, Canada (Marshall 1979). The most serious effects associated with conventional oil utilization stem from the emission of pollutants from transportation vehicles, which contributes significantly to the degradation of urban and regional air quality. Other significant impacts arise from offshore oil drilling and marine oil spills. Such spills—the majority of which originate not from catastrophic tanker accidents, but rather from routine, relatively low level discharges from ships flushing out empty tanks and from waste oil reaching the sea—can have drastic effects upon the ecology of coastal regions, polluting marshlands and estuaries and causing short- and long-term damage to marine organisms and aquatic and terrestrial wildlife.

The United States, as well as the other industrialized nations, faces a transition from a petroleum-dependent society to one that must rely on more abundant renewable energy sources. Planning of this transition requires a knowledge of the available time frame, making imperative an exhaustive, quantitative, and independent assessment of the magnitude of remaining petroleum reserves and of the potential rates and costs of future oil production. While this information has often been treated as the exclusive property of the oil companies, it is essential to the formulation of a rational policy concerning the prudent allocation of the petroleum resource.

NATURAL GAS

Natural gas is a clean-burning fuel whose major constituent is methane. Natural gas provides a major source of energy for the United States that, along with petroleum, has helped make possible the major expansion of the U.S. economy that has occurred since World War II. In contrast to petroleum, however, almost all of the natural gas consumed in the United States is domestically produced. Domestic supplies of natural gas are quite finite, and it is unlikely that production can be expanded significantly, unless large quantities of recoverable gas are discovered in unconventional sources. Therefore, the future use of this valuable fuel should be prudently and carefully administered.

In 1978, the United States consumed 20 trillion cubic feet (tcf), or about 20 quads,[3] of natural gas, which provided 25 percent of primary domestic energy use. Industry consumed about 42 percent of the total for use as a boiler fuel, a feedstock for the production of ammonia fertilizers, and numerous other applications. Roughly 39 percent of all natural gas was consumed in the residential and commercial sector mainly for building and water heating and for cooking. About 16 percent of all natural gas is burned to produce electricity, and the remaining 3 percent or so is consumed to power pumps for distribution in pipelines. At present, only about 5 percent, or 1 quad per year, of domestically consumed gas is imported, primarily from Canada.

Domestic reserves of natural gas are limited. At the end of 1978, proved natural gas reserves totaled 200 tcf (*New York Times* 1979) down from 290 tcf in 1970 (Hayes 1979). These discovered reserves would last only ten years at current consumption rates. Based on a historical recovery factor of 80 percent, undiscovered gas resources are estimated to lie in the range of 500 (Hayes 1979) to 1,000 tcf, including both conventional and unconventional sources (Mankin 1979). Total natural gas reserves and resources thus lie in the range of 700 to 1,200 quads. If all this gas could be economically produced—a prospect that is highly uncertain—it would be enough to provide a thirty-five to sixty year supply at current domestic consumption rates.

Estimates for total global natural gas resources are also quite uncertain. Proved reserves are estimated to be about 2,200 quads (*Oil and Gas Journal* 1977), while ultimately recoverable resources may range from 6,000 to 15,300 quads (Wilson 1977). Almost half of these reserves and resources are located in the Middle East and other developing nations, quite far from consuming centers. Traditionally, most of this gas has been flared off, but several gas-producing developing nations have ambitious plans to exploit it. Other nations, such as Algeria and Indonesia, are exporting natural gas in liquefied form. In the next decade, Mexico may become a large

---

3. One trillion cubic feet is equivalent to 1.035 quads.

exporter of natural gas to the United States. As much as 1.5 tcf per year may be piped across the border by 1988 (*National Journal* 1979).

Natural gas occurs in rock formations containing oil or similar to those containing oil. It may be found dissolved in the oil, unmixed in a cap above the oil, or by itself. About 80 percent of the gas in a reservoir is generally recovered, although according to the U.S. Geological Survey (USGS 1975a), up to 90 percent may be extracted by enhanced recovery methods. It may also be possible to extract natural gas from so-called "unconventional" sources. These include deep deposits—those below 25,000 feet—such as Appalachian shale, tight sandstones located in western states, coal seams, and the geopressurized brines of the Gulf Coast. At the present time, however, recovery technologies for these nonconventional sources are either in an early stage of development or nonexistent.

Historically, natural gas has been moved to consuming centers by pipeline. In the last decade, however, increasing amounts have been transported and stored as liquefied natural gas (LNG). "Peak-shaving" gas is liquefied in the summer time from domestic supplies and stored for peak use on cold winter days. This practice supplies up to 15 percent of the peak demand for natural gas in some areas of the northeastern United States. Imported natural gas is liquefied near the point of overseas production, transported in refrigerated tankers at a temperature of $-260°F$, and regasified and introduced into pipeline systems near points of consumption. The LNG importation system requires a large capital investment, has high transportation costs, suffers energy losses of about 25 percent in processing, and entails the possibility of large-scale accidents involving fire or even explosions.

The economics of natural gas and petroleum are in many ways similar. Production costs have traditionally been low, and since 1954, the price paid to natural gas producers by the interstate pipeline companies has been regulated by the federal government. These prices reflected the average rather than the marginal cost of gas production. By contrast, intrastate prices were unregulated and, in 1973, began to rise, reflecting, in part, the increased costs of finding new gas supplies. In recent years, interstate gas prices have been allowed to rise too, freeing supplies previously retained for intrastate sale. Pricing of newly discovered natural gas has been deregulated, and under the provisions of the National Energy Act of 1978, all natural gas pricing is to be freed from federal regulation by 1985. The net effect of deregulation on increased production of natural gas remains to be seen. As is true of petroleum, there appears to be an inevitable prospect for rising natural gas prices.

By comparison with other fossil fuels, natural gas is a benign energy source. Its combustion produces almost no by-products other than carbon dioxide, water vapor, and heat. However, serious safety problems are posed by the extreme fire hazard associated with the transport and storage of LNG in or near heavily populated areas. If leaking LNG were ignited following an accidental rupture of a large tanker flames could extend upward one mile high and cause injuries within a several mile radius from the spill site (Fay and MacKenzie 1976). A significant risk is also posed by truck transport of LNG through metropolitan areas. Thirteen LNG import terminals were planned for the United States by 1980; however in early 1979, two license applications were rejected by the Department of Energy, indicating a reappraisal of the role of LNG in the United States' energy future. In any event, it seems clear that for safety reasons, LNG terminals should be remotely sited from cities and populated areas, and the transportation of LNG by truck and rail through heavily populated areas should be prohibited, with reliance placed instead on gas distribution by pipeline.

Given its present importance and inherent attractiveness, natural gas will serve as an important "bridging" fuel in the transition to a renewable energy future. While the magnitude of the extractable natural gas resource is not precisely known, remaining supplies appear sufficient to supply current levels of consumption at least through the year 2000 without significant imports if a rational policy of natural gas production and use is implemented. Such a program might include the phaseout of natural gas use by industry and utilities, to be replaced by other energy sources and preferential allocation of gas to homes, small businesses, and commercial establishments. An informed policy would also depend on the availability of accurate data on gas reserves and potential production rates and costs in order to allocate remaining quantities of this fuel in an intelligent and judicious manner. Greatly increased use of natural gas would probably require an increased dependence upon imports of LNG. The safety and economic consequences of heavy reliance upon imported LNG must be thoroughly assessed, however, before such a course of action is adopted.

OIL SHALE

Oil shale is a very fine-grained sedimentary rock containing organic matter that can be converted into a heavy crude oil upon heating. It is a fossil fuel resource present in very great quantities in the United States, particularly in western parts of the country. However, for

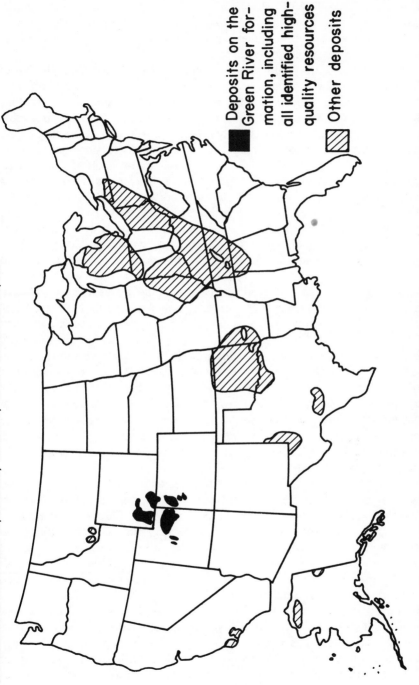

Map 3-3: Oil Shale Deposits in the United States. The Green River formations contain about two trillion barrels of oil, while the eastern and midwestern deposits may contain up to two trillion barrels, too.

Deposits on the Green River formation, including all identified high-quality resources

Other deposits

Source: Duncan, D.C., and V.E. Swanson. "Organic-Rich Shale of the United States and World Land Areas." U.S. Geological Survey Circular 523. Washington, D.C.: Government Printing Office, 1965.

economic reasons, oil shale has not yet contributed appreciable amounts of energy to the U.S. supply. The future development of this resource will be impeded by a combination of high production costs and severe environmental impacts.

The richest oil shale deposits in the United States are found in the Green River formations in Colorado, Wyoming, and Utah. The total identified resource in this region is the equivalent of 2 trillion barrels of oil (Maugh 1977), or 11,000 quads. If recovered, this could ultimately provide the equivalent of almost 300 years of oil consumption at current domestic rates. The total undiscovered resource may be more than a factor of ten higher than the identified resource (Smith and Jensen 1976). Eastern and midwestern oil shales may contain up to another 2 trillion barrels of oil (Maugh 1977). Despite these numbers, it is estimated that only about 80 billion barrels (about 450 quads) are recoverable using conventional technology (Bureau of Mines 1975). With the development of new technologies, the recoverable resource could possibly run as high as 200 billion barrels, or 1,200 quads. At present it is difficult to predict how large the recoverable resource might be because this quantity is intimately tied to the uncertain economics of oil shale production.

Oil shale typically contains about 15-20 percent organic matter. Of this, about 10 percent is insoluble kerogen, which provides the binder for the shale, and the remaining 90 percent is a soluble material called bitumen. When the shale is heated to 900°F in the absence of oxygen, the organic matter melts. At higher temperatures, gases are given off. Some 25-75 percent of the organic matter can be converted to oil and combustible gases; the remainder forms combustible char and coke. After retorting (the process in which the shale is crushed and heated), the shale oil is upgraded by hydrogenation, making it suitable for refining.

Three different technologies for recovery of oil shales are under development. The first—mining and surface retorting—involves extraction of shale by conventional mining techniques, followed by processing in a surface facility. The second—*in situ*, or in place, recovery—involves underground fracturing and ignition of the crushed shale. As the flame front advances through the rubble, oil is released and pumped to the surface. The third technology—modified in situ recovery—is a combination of the first two methods. A shallow room is mined out of the shale, the roof is blasted down, and the rubble ignited. The mined shale is processed on the surface. Neither of the *in situ* processes has yet been demonstrated on a commercial scale. Surface retorting has been commercially utilized in the past in Europe and the Appalachian region, but was not economically com-

Photograph 3-2: Oil Shale Retort. This pilot plant, located at the Colony Development Operation facility, Parachute Creek. Colorado, processed 1,000 tons of oil shale per day from 1969 to 1972.

Source: Colony Development Operation, Atlantic-Richfield Company.

petitive with liquid petroleum and was thus phased out. Surface retorting, although a well-developed technology, is subject to major environmental problems as discussed below.

A practicable oil shale industry must be economically competitive not only with newly emerging renewable energy sources but also with liquid petroleum. At times, as the price of petroleum has risen, it has seemed as though shale oil might achieve competitiveness, but invariably, the projected price of shale oil has also increased and has eliminated this hopeful prospect. Mining and processing of oil shales are highly capital-intensive enterprises particularly susceptible to inflation. For example, in 1973 the Colony Development Corporation proposed construction of a 45,000 barrel per day surface retorting plant, estimating its cost to be about $250 million. By 1977, the cost estimate had climbed to $1.2 billion (Maugh 1977). At the present time, the cost of a barrel of shale oil is estimated to lie in the range of $25-35 (White House 1979). However, no one yet knows precisely how much it will cost to produce a barrel of oil from shale. According to a Rand Corporation study prepared for the Department of Energy, cost overruns by a factor of about three are not at all unlikely for synthetic fuel plants (Cowan 1979). The economic competiveness of shale oil would surely be enhanced if Congress authorizes the $3 per barrel tax credit to producers proposed by President Carter in his July 1979 address on national energy policy. It is also possible that new technological processes under development that yield alumina and soda ash—both readily saleable—in addition to oil could do much to alleviate the presently unfavorable economics (Maugh 1977).

Shale oil is not a clean-burning fuel. Compared to liquid petroleum, it contains high concentrations of nitrogen, sulfur, and other potential pollutants. Surface recovery processes would significantly affect air quality and degrade large areas of land with impacts comparable to those from the strip mining of coal. A 1 million barrel per day surface retorting operation could affect up to 450,000 acres (700 square miles) of land over a thirty-year period. Both surface and modified *in situ* retorting produce large quantities of spent shale waste, and processing increases the volume of the shale so that not all of it can be returned to the mine. The scale of oil shale processing can be appreciated by noting that it takes anywhere from 1 to 3 tons of shale to produce a single 42 gallon barrel of oil (Maugh 1977). Production by means of surface retorting would involve the mining and processing of about 7 tons of rock and overburden on the average to produce the energy equivalent of 1 ton of coal. The large volume of wastes left over from this process would have to be disposed of on the surface or in canyons, and the leaching of salt and other minerals from these dumps would cause significant pollution of water supplies. Surface retorting consumes 2 to 5 gallons, and the modified in situ process requires 1 to 3 gallons of water for each gallon of oil produced. The *in situ* process requires less water and has a less severe environmental impact. The water requirements of a large-scale oil shale industry could cause severe problems in the arid western states with their limited water resources.

The potential resources to be obtained from oil shale are, quite literally, enormous and could ultimately extend domestic petroleum supplies for many decades. However, the environmental impacts of oil shale recovery are quite significant. Furthermore, the dubious economics of shale oil production suggest, at least for the forseeable future, that oil shale will be a useful resource only if efforts to develop renewable energy alternatives are not forthcoming.

NUCLEAR FISSION—LIGHT WATER REACTORS

Nuclear fission is a process that allows for the exploitation of the enormous energies locked in the atomic nucleus. The complete fissioning of one pound of uranium-235[4]—the only naturally occurring fissile isotope of uranium—yields the energy equivalent of about 1,000 tons of coal. While great expectations were previously placed on this power source, with visions of 1,000 or more reactors operating by the turn of the century, they most likely will never materialize. After nearly three decades of government-subsidized support, nuclear power still remains a small contributor to the nation's energy supply. Future prospects are clouded by a number of broad-ranging problems—sharply escalating plant construction costs; a growing recognition of widespread safety deficiencies affecting operating plants; the lack of a demonstrated and safe means of disposing of radioactive wastes; and mounting concern over the threat of nuclear weapons proliferation.

In 1978, nuclear fission reactors generated approximately 13 percent of the total electricity consumption in the United States, equal to about 3 quads of primary energy or less than 4 percent of total gross energy use. At the beginning of 1979, there were seventy commercially licensed plants with an aggregate installed generating capacity of about 51,000 MWe. An additional 126 plants, with a total capacity of 140,000 MWe, were either under construction or on order.

Nuclear technology is based upon the principle whereby a fissile atom, such as uranium-235 or plutonium produced only by reactor operation, when struck by a low energy, or "thermal," neutron, may split into fragments, or "fission products," yielding neutrons and a

4. Uranium consists primarily of two isotopes—uranium-235 and 238. U-238 is by far the more prevalent isotope, comprising about 99.3 percent of natural uranium, with U-235 accounting for most of the remaining 0.7 percent. Each isotope contains the same number of protons—92—but differing numbers of neutrons—143 in uranium-235 and 146 in uranium-238. Only U-235 is readily fissionable in LWRs.

Map 3–4: Nuclear Power Reactors in the United States.

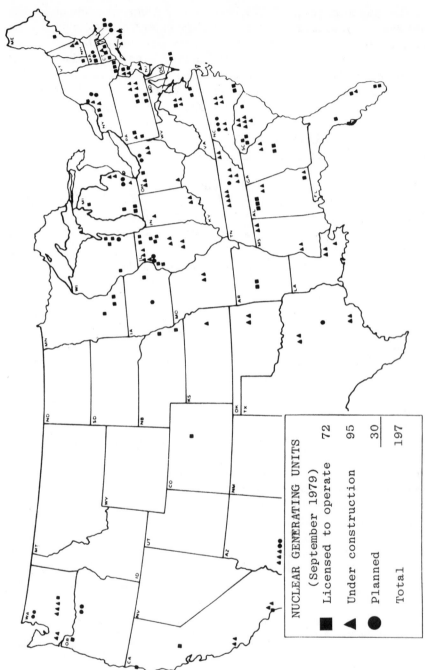

NUCLEAR GENERATING UNITS
(September 1979)

■	Licensed to operate	72
▲	Under construction	95
●	Planned	30
	Total	197

Source:: Department of Energy. "Nuclear Reactors Built, Being Built, or Planned in the United States as of Dec. 31, 1978."
TID-8200-R39 Oak Ridge Tenn.: Technical Information Center, March 1979, p. 4. (Updated through September 1979.)

concomitant energy release. Although a number of other fission reactions exist, some of which could be utilized in nuclear reactors,
the current nuclear fission program of the United States is based
solely on the utilization of uranium resources in so-called light water
reactors (LWRs). For this reason, alternative fission cycles are not
discussed here in detail.

Economically exploitable high grade uranium resources suitable
for LWR use are limited, and therefore, the energy potential of the
current generations of LWRs is not very great. Estimates of recoverable uranium resources suggest that between 700 quads, assuming no
recycling of fissile nuclides, and 2,400 quads, assuming uranium and
plutonium recycle, could be made available (ERDA 1976b; DOE
1978)—sufficient resources to fuel somewhere between 400 and
1,400 large, 1,000 MWe LWRs operating at an average 65 percent
capacity factor, each over a typical lifetime of thirty years. It should
be noted, however, that advanced fission converter reactors, based on
uranium-thorium recycling, could, theoretically, increase substantially
the upper limit on the total energy available (Feiveson, von Hippel,
and Williams 1979).

The basic purpose of a fission power reactor is to boil water. The
steam is used, as in a coal- or oil-fired plant, to generate electricity.
The light (ordinary) water circulating through the reactor core, where
the fission process takes place in the uranium fuel, draws off heat.
Depending upon reactor type—there are two currently in use in the
United States, the boiling water reactor and the pressurized water
reactor—the cooling water is then either converted directly into steam
or used to heat water in a secondary cooling loop that boils to produce steam. In either case, the steam is used to drive a turbine generator (Glasstone 1967).

The entire process of producing energy from uranium is a sequence
of many complex operations called the "uranium fuel cycle." This
cycle consists of mining and milling of uranium ore; enrichment of
uranium-235 content from the natural 0.7 percent to the 3–4 percent
required for LWR operation; fuel fabriction; fissioning of uranium in
a reactor; reprocessing of spent fuel and recycling of fissile material
(remaining uranium-235 and plutonium-239 produced in the reactor)
into new fuel; or disposal of spent fuel and radioactive waste. Not all
parts of the fuel cycle have been fully developed. Reprocessing and
recycling have not yet achieved commercial success and, indeed, may
never do so. They have, in any case, been indefinitely deferred by
presidential decision. Long-term disposal of spent fuel and radioactive waste will not be initiated before 1990 and possibly later (UCS
1975; Metz 1977; IRG 1979; Lipschutz 1980).

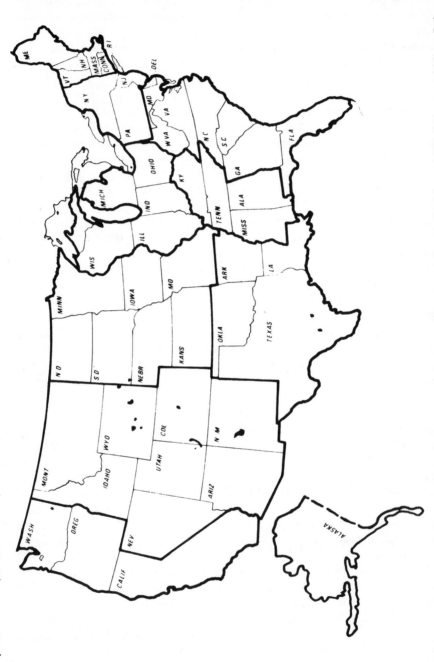

Map 3-5:   Principal Uranium Deposits of the United States.

Source: Crump, L.H. "Fuels and Energy Data: United States by States and Census Divisions, 1974." Bureau of Mines Information Circular 8739. Washington, D.C.: Government Printing Office, 1977. p. 12.

Contrary to early predictions of "power too cheap to meter," nuclear-generated electricity, while cheaper than that from oil-fired plants, has become comparable in cost to electricity from coal plants. Power costs from some reactors built years ago have been low for several reasons. Government subsidies have helped to keep costs artifically low in the past and continue to do so. A number of early commercial reactors were sold as "turnkey" operations at fixed cost by manufacturers who then absorbed any losses. Early reactors were smaller and employed cheaper, less sophisticated safety systems. Up-graded safety standards have added to the expense of plant construction and operation and will continue to do so in the future. Cost overruns for nuclear plants have been enormous, typically in the range of 100 to 200 percent. It presently costs about 50 percent more to build a nuclear plant than a coal plant: nuclear plants finished today cost about $1,000 per installed kilowatt of capacity, whereas coal plants, employing the best available pollution control technology, average about $675 per kilowatt (Parisi 1979b). Over recent years, the cost of building nuclear plants has been increasing by about 26 percent annually, while construction costs for coal plants have been increasing by only about 17 percent per year. By the mid- to late 1980s, nuclear plants may cost twice as much to build per kilowatt as coal-fired generating plants (*Nucleonics Week* 1979). Other parts of the fuel cycle, most notably the cost of uranium ore, have been subject to even larger rises in costs.

The environmental and health hazards of nuclear power result primarily from the radioactivity associated with various parts of the fuel cycle. Under normal operating conditions, radiation releases from nuclear plants are small, and the health hazard to the public is minimal (less, in fact, than that associated with coal-fired plants). However, the great radioactive inventory of a nuclear plant requires extreme vigilance to ensure that none or very little is allowed to escape. Despite the presence of sophisticated safety systems, there are many unresolved safety problems in operating nuclear plants. As events in March 1979 at the Three Mile Island nuclear generating plant near Harrisburg, Pennsylvania, demonstrated, unanticipated conditions within the reactor can lead to the release of radiation that may threaten the health and safety of the public. Catastrophic re-leases of radiation from a disabled reactor, by no means an im-plausible occurrence, could result in tens of thousands of deaths and hundreds of thousands of injuries (NRC 1975; UCS 1977). Radiation hazards from mining, milling, reprocessing, and waste disposal may also result in measurable detrimental effects to the health of workers and the public (Lipschutz 1980).

Photograph 3-3: Nuclear Fission Generating Station. Aerial view of the Indian Point 2 and 3 nuclear generating plants, operated by Consolidated Edison and the New York Power Authority, respectively. The plants are pressurized water reactors. Unit 2 is rated at 873 megawatts; Unit 3, 965 megawatts. Unit 1 (low dome at right) is permanently out of service.

Source: Henry W. Kendall.

Finally, the potential for nuclear weapons proliferation as a result of large-scale implementation of reactor and reprocessing technology cannot be ignored. As demonstrated by the Indian atomic explosion in 1974, weapons technology is well within the grasp of most nations possessing some type of research or commercial reactor program.

Unresolved problems relating to reactor safety, radioactive waste disposal, and nuclear weapons proliferation make an increased dependence upon nuclear energy technology an imprudent course for the United States at the present time. Moreover, the economic and environmental advantages of nuclear power are by no means clear. Furthermore, light water fission reactors cannot provide great amounts of energy owing to the limited quantity of high grade domestic uranium resources. Significant expansion of the energy potential of nuclear fission would require resorting to advanced converter reactors based upon uranium-thorium recycling. Even greater expansion of energy supplies could be obtained from the breeder reactor—a technology with certain inherent advantages, but some very grave disadvantages. This is discussed in the following section.

NUCLEAR FISSION—THE BREEDER REACTOR

A fission breeder reactor is a device that creates more fissionable material than it consumes, while simultaneously producing electricity. It does this by means of nuclear reactions induced in non-fissile, "fertile" material placed in the reactor, which converts them to fissionable materials. Uranium-238, which comprises 99.3 percent of naturally occurring uranium, will not fission in a conventional reactor but can be used in a breeder. Because potential resources of fertile material—uranium-238 and thorium-232—are quite large, the breeder can be viewed as a long-term energy source. Although fission breeder power reactors are under development in a number of countries, the current role of the breeder in energy production worldwide is still extremely limited. In fact, in the United States, breeder reactors will not be producing electricity on an extensive basis during the remainder of this century. Furthermore, there are potentially adverse social, environmental, and economic consequences associated with a large-scale breeder program that, in our view, cast strong doubt on the wisdom of ever relying on this particular means of electricity generation.

A breeder reactor could produce fissile material either from fertile uranium-238, thereby producing plutonium-239, or from thorium-232, yielding uranium-233. In either case, an initial charge of fissile material—uranium-235 or plutonium-239 produced in thermal fission reactors—would be needed to start up the breeder. A number of uranium- and thorium-based breeder reactors have been proposed, such as the light water breeder (using ordinary water as coolant), the gas-cooled fast reactor (employing helium as coolant), and the molten salt breeder (in which the liquid coolant also contains fissile and fertile material). (Glasstone 1967; Willrich and Taylor 1974). The breeder reactor under the most intense developmental effort is the liquid metal fast breeder reactor (LMFBR), which transmutes uranium into plutonium and has a liquid sodium coolant. For this reason, the discussion here will focus almost exclusively on this type of breeder.

The energy potential of the LMFBR is quite large because it is theoretically capable of utilizing uranium resources some sixty times more efficiently than present-day light water reactors. Assuming that only high grade uranium oxide reserves and probable resources are exploited, between 40,000 and 100,000 quads (ERDA 1976b; DOE 1978) of primary energy in the form of electricity could be produced, a 500 to 1,300 year supply of energy at current domestic rates of gross energy use. Utilization of low grade uranium resources could extend the upper limit to as much a 1 million quads, although excessive environmental costs would be likely to restrict exploitation of these resources.

The process of nuclear fission in the LMFBR is essentially identical to that in an LWR. The principal fissile element in the reactor fuel is plutonium-239. Large amounts of fertile uranium-238 are placed around the reactor core in a kind of "blanket." The uranium is converted to plutonium by the process of neutron capture. The plutonium in the core is fissioned by high energy, or "fast," neutrons, producing fission fragments and more fast neutrons. Heat produced in the fuel is transferred to the liquid sodium coolant. The coolant circulates through heat exchangers, producing steam that is used to generate electricity. Some of the neutrons, left over and unneeded to sustain the chain reactions, will create plutonium in the blanket. Under favorable conditions, the process will produce more fissile material than it consumes, as long as a supply of fertile material is available. Present-day breeders may require as much as some thirty-five to sixty years to produce as much plutonium as is consumed, but reactor engineers hope that newer designs may be able to shorten this "doubling time." Decreasing the doubling time requires that the breeder core be made very compact so as to increase the fast neutron density. However, a compact core and high neutron density tend to make the breeder more difficult to control and therefore more vulnerable to major instability and so, in turn, to catastrophic accident.

As has been true of the light water reactor program, the LMFBR program has been subject to vast cost overruns. The estimated cost of the prototype breeder, planned for Clinch River, Tennessee, has escalated from $700 million in 1972 to well over $2 billion in 1978, equivalent to a capital cost in excess of $5,300 per kilowatt. Studies suggest that breeder capital costs will exceed those of a comparably sized LWR by 25–75 percent (Feiveson, von Hippel, and Williams 1979). Furthermore, expected costs for the breeder fuel cycle have risen to the point where significant savings from the more efficient use of uranium resources can no longer be depended upon.

Although all of the environmental, health, and safety risks of the LWR are also applicable to the LMFBR, there are several problems of special concern that are unique to the breeder. The presence of large amounts of fissile material at a high density within the core could lead to runaway criticality and even a very low magnitude nuclear explosion in the event of insufficient control over the chain reaction or a fuel meltdown causing rearrangement of fuel geometry. Such an event could cause rupture of the containment structures and release very large amounts of radioactivity into the environment. The liquid sodium coolant in the LMFBR is also highly reactive. It burns in air and explodes upon contact with water and so can be the initiator of a major upset.

Perhaps the most serious problems would be those associated with the large inventories of plutonium contained in the fuel cycle of a breeder-based system. Plutonium is highly toxic if inhaled in particulate form (Bair and Thompson 1974). Even an extremely small loss through ordinary processing and handling of the large quantities of this material circulating in a breeder economy could pose serious health hazards to the public. More threatening would be the intentional diversion of small quantities of plutonium for the purpose of fabricating a nuclear weapon. A low yield nuclear device would require only about twenty pounds of reactor grade plutonium. The security requirements needed to ensure that such diversions would not take place would be extensive and pervasive, and major infringements of civil liberties in order to safeguard the fuel cycle would not be implausible (Ayres 1975). The concentrated siting of up to twenty-five large breeder reactors, along with reprocessing and waste management facilities, in so-called "breeder parks," as suggested by some in response to the security question (Weinberg 1976), would surely create problems of ensuring safe and responsive administration. Furthermore, a large release of radiation from one plant could require the shutdown of a major portion of the entire park, causing massive disruptions of service and possibly leading to accidents as neighboring plants were abandoned.

Alternate breeder cycles, based upon use of thorium-232, have also been suggested as a "solution" to the proliferation problem, because the fissile uranium-233 produced in such a cycle could only be extracted by technically demanding isotopic separation techniques rather than by the simpler chemical processes required to separate plutonium from uranium. However, persuasive arguments suggest that a thorium-based cycle would offer little, if any, proliferation-resistant advantages (Lovins 1979). In any case, the initial charge of fuel to a thorium breeder would have to contain significant quantities of uranium-235 or plutonium-239, both of which could be extracted by terrorists for use in nuclear weapons.

Viewed in terms of economic, social, and environmental costs, the choice of whether or not to proceed with a large-scale breeder program must be made with great care and deliberation. Because of its efficient utilization of uranium resources, the breeder could supply a substantial portion of our energy requirements over an extended period of time. The outstanding problems regarding safety, health hazards, and costs, however, suggest that the demanding responsibilities associated with the breeder option may be greater than society is satisfactorily prepared to handle. In any event, a decision to proceed need not, and we believe should not, be made immediately. It is clear that breeder technology must be regarded as a last resort energy source. It is a conclusion of our report that breeder technology need never be implemented if the country pursues an intelligent course of energy development.

NUCLEAR FUSION

Nuclear fusion occurs when two light atomic nuclei, such as hydrogen, approach closely enough to overcome their mutual electrostatic repulsion. The attractive nuclear force causes them to fuse together and release energy. Fusion is the process that produces energy in the sun. It is uncontrolled fusion that is the source of energy for thermonuclear weapons, hydrogen bombs. Twenty-five years of earthbound research into controlled fusion, however, have yet to produce a technology able to yield useful energy. To a large degree, this research has instead served only to underline the difficult nature of the technological problem. The feasibility of fusion energy has not yet been demonstrated. It is clear that the commercialization of controlled fusion cannot occur until the next century, if ever.

Several fusion reactions are of particular interest to controlled fusion research. Each has some advantages and some major problems. A final choice has not yet been made. The deuterium-deuterium (D-D) and deuterium-tritium (D-T) reactions utilize heavy isotopes of hydrogen,[5] but both produce energetic neutrons that would render a fusion device highly radioactive. The proton-boron-11 reaction could also be of interest, particularly because it produces charged particles, rather than neutrons, which would not induce radioactivity in a reactor.

Deuterium is quite common in sea water, and the deuterium content of the world's oceans would be sufficient to sustain a level of energy consumption equal to present world consumption, if based on the D-D reaction, for thirty billion years (Holdren 1978). Tritium,

---

5. While an ordinary hydrogen nucleus has only one proton, deuterium has a neutron in addition to a proton, and tritium has two neutrons plus one proton.

however, is a radioactive isotope with a twelve year half-life and does not exist naturally. It must be "bred" from lithium, and for this reason, lithium is the limiting factor in determining the energy potential of the D-T reaction. (In this sense, the D-T fusion reactor would be a "breeder.") Boron resources are quite large, and protons can be obtained from water. Thus, if controlled fusion can be achieved, potential energy supplies will be extremely large. A low estimate of potential energy yield, based on D-T fusion and constrained by domestic supplies of lithium, is 200,000 quads (Holdren 1978). A high estimate, based on proton-boron-11 fusion is 1 billion quads. At current rates of domestic energy consumption, this would roughly provide a 2,500 to 10 million year supply. Sufficient global resources of lithium exist to raise the lower bound estimate considerably.

An important constraint on choosing the appropriate reaction is set by the physical conditions required to achieve fusion. Unlike nuclear fission, which can take place at room temperature, fusion reactions can occur only at extremely high temperatures, in the range of about 200 million to 20 billion degrees Fahrenheit. The D-T reaction falls at the lower end of this range and is thus considered relatively "easy" to achieve. D-D fusion requires energies in the intermediate range, while the proton-boron-11 reaction falls at the upper end. Thus, current research is centered on the D-T reaction.

The extreme temperature conditions have necessitated development of novel techniques to confine the reaction materials, which, when reacting, are in the form of hot, rarified, electrically conducting gases called "plasmas." No ordinary methods will work. To meet this problem, two containment techniques have been developed. One, magnetic confinement, involves confining the hot plasma within a suitably shaped magnetic field. The other, intertial confinement, involves the almost instantaneous compression of a D-T fuel pellet by a bank of laser or particle beams. Both approaches seek to create the conditions necessary for successful fusion. However, research to date has failed to achieve the energy "breakeven" point at which the input energy needed to induce fusion in a plasma or pellet is equal to the energy produced by the fusion reaction (Metz 1976b; Post 1976; Rose and Feirtag 1976). A number of experimental devices that may achieve breakeven are planned for operation in the early to mid-1980s (Stickley 1978; Yonas 1978).

A fusion reactor will consist of a reaction chamber surrounded by a liquid lithium blanket. High energy neutrons produced in the chamber will pass through a wall into the lithium, losing energy and producing tritium. The lithium will be circulated through a steam

Photograph 3-4: Experimental Inertial Confinement Fusion Device. Target chamber of the Shiva inertial confinement fusion experiment at Lawrence Livermore Laboratory. The chamber has portholes for the twenty laser beams that will compress a microscopic target fueled with a deuterium-tritium mixture, plus other portholes for measurement and observation instruments.

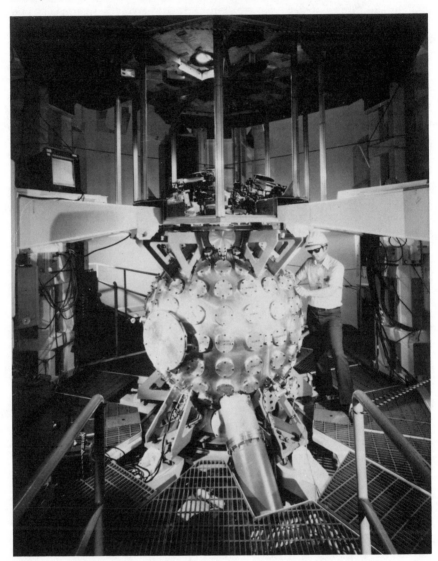

Source: Lawrence Livermore Laboratory.

generator, and the steam will be used to generate electricity. In the past, government schedules called for the commmercialization of fusion by the year 2000, an unrealistic goal. Current projections suggest that a prototype fusion power reactor may be developed by 2005, provided scientific feasibility can be achieved, but commercialization will not be reached before 2020 at the earliest (Sullivan 1979).

An alternative approach to pure fusion that could shorten the time required for commercialization is the fusion-fission hybrid concept. A reactor of this type would operate at less than breakeven, producing some power by fusion while breeding fissile plutonium from uranium-238. This concept appears attractive to some, because the D-T reaction produces copious quantities of the high energy neutrons that are required to produce fissile plutonium. However, the hybrid would be subject to serious problems: it would require nuclear fuel reprocessing, would be technologically very complex, and would lose many of the inherent safety and environmental advantages that fusion power might hold over nuclear fission. In a sense it can be considered to combine some of the worst features both of fusion and of fission.

Research and development costs for a fusion reactor will be high, requiring up to $10 billion for the development of a 10 megawatt electric (MWe) prototype (Metz 1976a). By contrast, a 20 MWe fission breeder prototype cost only $32 million, while the estimated cost for a 10 MWe solar electric generating prototype is $90 million. But at this early date, all cost estimates for a commercial fusion power reactor must be considered highly speculative. Unless a simple approach to fusion is developed, a fusion power reactor is likely to cost at least as much, and possibly three to five times the cost of a comparable fast breeder fission reactor (Metz 1976a). The fusion-fission hybrid is also almost certain to cost more than a "pure" fission breeder (Metz 1976c), if indeed it works at all.

Although the D-T reaction appears to be the easiest to achieve technically, it is subject to several environmental and safety problems. It utilizes radioactive tritium, which is difficult to contain. Accidental releases of tritium could pose a health hazard, but the magnitude of potential releases would be much less than that resulting from a catastrophic fission reactor accident. Furthermore, fusion reactors will be, overall, less dangerous than fission reactors because the possibilities of runaway criticality in the breeder and of core meltdown in all fission reactors are absent. The high energy neutrons produced by the D-T reaction would cause radioactive activation of reactor components and, ultimately, a radioactive waste management problem as periodic replacement of reactor components produced radioactively contaminated materials. Use of exotic alloys, such as vanadium, could limit the waste problem. By contrast, the proton-boron-11 reaction, although much more difficult, if not impossible, to achieve, would require no tritium and produce no radioactive wastes.

Another serious problem associated with fusion is the link with nuclear weapons. Any fusion reactor running on a cycle that produces a high neutron flux could be used to create fissile materials suitable for use in fission weapons. In particular, a hybrid reactor would give rise to the same proliferation concerns that are associated with the fission breeder reactor.

Fusion could supply a virtually limitless amount of energy, assuming constraints on the availability of critical materials for reactor construction could be avoided. Given the current pace of research and development, however, fusion is not likely to achieve commercialization until well into the twenty-first century. At that time, fusion may be unnecessary if renewable energy alternatives—some already close to commercialization—are then supplying a large portion of the United States' energy needs.

GEOTHERMAL ENERGY

Geothermal energy embodies the vast reservoirs of heat contained in the earth's crust. If this heat could be fully tapped, it would provide an essentially limitless source of energy. However, practical considerations, the lack of appropriate technology, and possibly severe environmental impacts make large-scale exploitation of this energy source impossible, at least in the near future. Small-scale utilization of geothermal energy has been and will continue to be possible in the United States on a strictly regional basis. In 1978, geothermal energy supplied less than 0.1 percent of the energy consumed in the United States, primarily in the Geysers basin area of California. In the long term, it may be possible to produce amounts of energy significant to the country's supply from geothermal sources, if technological and economic hurdles can be overcome.

Although the potential geothermal resource is truly prodigious, it must be considered nonrenewable. The rate of energy flow from within the earth's interior toward the surface is extremely limited. While certain deposits of stored geothermal energy can be exploited, they will only be regenerated very slowly, making the resource, for all practical purposes, inherently depletable.

Geothermal resources fall into four general categories—hydrothermal convection zones, geopressurized fluid zones, hot rock, and

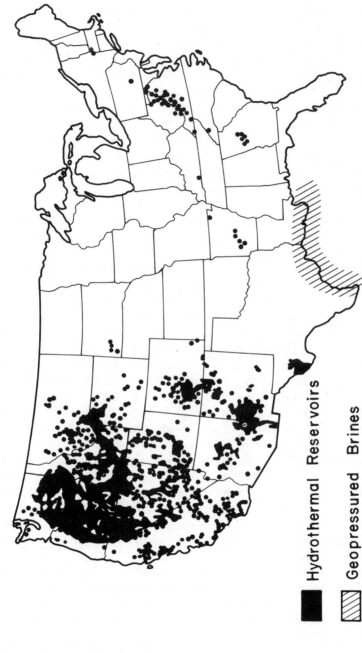

Map 3-6: Location of Geothermal Resources in the United States.

■ **Hydrothermal Reservoirs**

▨ **Geopressured Brines**

Source: Department of the Interior. "Final Environmental Statement for the Geothermal Leasing Office." Washington, D.C.: Government Printing Office, 1973. Vol. 1, p. II–17.

magma. Hydrothermal zones, such as can be found in some western states, are regions of hot rock overlying magma domes that contain large amounts of groundwater. The water circulates between the heat source and the surface, emerging either as vapor or liquid. Geopressurized fluid zones are geologic formations in which water is trapped at very high pressures between layers of impermeable shale. It is believed that these fluids may contain large amounts of methane, itself a valuable energy source. Such areas can be found extensively along the Texas and Louisiana Gulf coasts, but the precise quantities of methane there and the fraction that is recoverable are entirely speculative at present. Hot dry rock, with temperatures as high as 600° F, is usually found deep in the earth, where there is little or no groundwater. Magma systems consist of molten rock, at temperatures in excess of 1,200° F, that appears close to the surface in areas associated with recent or ongoing volcanic activity.

Although the total stored thermal energy in the earth's crust is very large, only a small fraction is potentially extractable using existing or planned technologies. The simplest and cheapest way to use geothermal sources is in the form of direct thermal energy for home heating and industrial process applications. In hydrothermal convection systems, deep underground water is heated under pressure. Brought to the surface, it may be utilized either as steam or as liquid. In vapor-dominated areas, such as the California Geysers basin, steam is used to power electric turbine generators directly. In liquid-dominated areas, such as exist in Iceland, hot water for heating is piped directly to homes. Although the technology for tapping geopressurized zones does not yet exist, it is believed that suitable extraction methods can be developed from deep drilling techniques pioneered by the oil industry. Extraction of thermal energy from hot dry rock would require the drilling of deep wells and fracturing of the rock by explosive or hydraulic techniques. Water would be injected into one well, passed through the fractured zone and heated, and withdrawn from a second well. A pilot hot dry rock project is currently underway near Los Alamos, New Mexico, but the technology is still in an early stage of development. At the present time, no practical means of exploiting magma systems have been developed or proposed.

Hydrothermal convection and geopressurized fluid zones are considered exploitable using existing or foreseeable technology. Potentially extractable geothermal energy in the United States is estimated to lie between 300 quads (electricity, 140 quads; thermal energy, 60; methane from geopressurized fluids, 100) and 1,000 quads (electricity, 410; thermal, 350; methane, 240) (USGS 1975b), although

Photograph 3-5:   Geothermal Energy. Unit 3 and 4 of Pacific Gas and Electric's generating units at The Geysers in northern California. The total capacity of the 12 generating units at the power complex is 608 megawatts.

Source: Pacific Gas and Electric Company.

these numbers are subject to downward revision if geopressurized resources prove difficult or impossible to exploit. In particular, the quantities of methane that can be extracted from geopressurized fluids may be a good deal less than suggested here. If technologies to exploit hot dry rock can be developed, however, these estimates might be increased considerably.

Proposed geothermal-based electrical generation systems may be economically competitive with coal and nuclear power in the near future. At the present time, however, the only competitive system in the United States is the one operating in the Geysers basin dry steam region, where electrical generation costs are comparable to coal and nuclear power and lower than oil generation. Capital costs for hydrothermal convection systems in areas with less easily exploitable sources than the Geysers basin are very likely to be higher. The costs of geopressurized electrical generation and methane extraction systems are expected to be substantially higher than for conventional energy sources, but in the absence of demonstrated technologies for exploiting geopressurized sources, cost estimates are speculative at best.

Geothermal energy systems may have significant environmental impacts, including air, water, noise, and thermal pollution; heavy land use; and seismic disturbances. Some geothermal fluids contain large quantities of noxious gases and high concentrations of salts and toxic metals that can significantly pollute air and water. Geothermal energy production systems are often quite noisy and, because of low conversion efficiencies, reject a large amount of thermal energy to the environment. Geothermal energy production involves the commitment of large land areas. The removal of large quantities of underground fluid can result in serious land subsidence, as has been noted along the Texas Gulf coast. Extensive exploitation of geopressurized reservoirs could result in subsidence exceeding thirty-three feet in some instances (Kreitler and Gustavson 1976). Hydraulic fracturing of hot rock and injection of water into geothermal zones could cause slippage along existing geologic faults, a phenomenon noted with waste liquid injection wells and enhanced oil recovery operations in western states (*Mechanical Engineering* 1976). Because most potential hot rock areas are located in seismically active regions, the possibility of inducing earthquakes must be further investigated before extensive development of this resource is initiated.

The overall contribution that geothermal energy sources can make to the nation's energy requirements is limited, although it may be substantial in some regions of the United States. Large-scale exploitation of this resource must, however, await the development of appropriate technology as well as a thorough assessment of potential environmental impacts. In particular, where severe environmental impacts appear unavoidable, as in the case of land subsidence resulting from extraction of geopressurized fluids, geothermal energy sources should remain undeveloped.

REFERENCES

Ayres, R.W. 1975. "Policing Plutonium: The Civil Liberties Fallout." *Harvard Civil Rights Civil Liberties Law Review* 10, no. 2 (Spring) 369.

Bair, W.J., and R.C. Thompson. 1974. "Plutonium: Biomedical Research." *Science* 183 (February 22):715.

Bureau of Mines. 1975. "Shale Oil." In *Mineral Facts and Problems*. Bulletin 667. Washington, D.C.: Department of the Interior.

———. 1979. "Mining Assessed as Potential Method for Petroleum Production." Washington, D.C.: Department of the Interior, News Release, August 2.

Council on Environmental Quality (CEQ). 1978. "Environmental Quality. The Ninth Annual Report of the Council on Environmental Quality." Washington, D.C., December.

Cowan, E. 1979. "Synthetic Fuel Costs Called High." *New York Times,* July 11.

Cummings, R.G., et al. 1979. "Mining Earth's Heat: Hot Dry Rock Geothermal Energy." *Technology Review,* (February):58-78.

Department of Energy (DOE). 1978. "Statistical Data of the Uranium Industry." GJO-100. Grand Junction, Colo.

———. 1979. "Monthly Energy Review." DOE/EIA/0035-7 (79). Washington, D.C.: Energy Information Administration, July.

Dinneen, G.U., and L. Cook. 1972. "Oil Shale and the Energy Crisis." ASME Paper no. 72-WA/Fu-3. New York: American Society of Mechanical Engineers.

Energy Research and Development Administration (ERDA). 1976a. "A National Plan for Energy Research, Development, and Demonstration: Creating Energy Choices for the Future, 1976." ERDA-76-1. Washington, D.C.

———. 1976b. "ERDA Issues Latest Estimate of Uranium Reserves." ERDA News Release No. 76-94. Washington, D.C.

Fay, J.A., and J.J. MacKenzie. 1976. "Comments on the Safety Aspects of the Everett (Mass.) LNG Import Terminal." Cambridge, Mass.: Union of Concerned Scientists, October 8.

Feiveson, H.A.; F. von Hippel; and R.H. Williams. 1979. "Fission Power: An Evolutionary Strategy." *Science* 203 (January 26):330.

Glasstone, S. 1967. *Sourcebook on Atomic Energy.* 3rd ed. New York: Van Nostrand Reinhold Company.

Hammond, A.L. 1976. "Coal Research (IV): Direct Combustion Lags Its Potential." *Science* 194 (October 8):172.

Hayes, E.T. 1979. "Energy Resources Available to the United States, 1985 to 2000." *Science* 203 (January 19):233-39.

Holdren, J.P. 1978. "Fusion Energy in Context: Its Fitness for the Long Term." *Science* 200 (April 14):168-80.

Hubbert, M.K. 1974. "U.S. Energy Resources, A Review as of 1972." A National Fuels and Energy Policy Study. 93rd Cong., 2nd sess. Senate Committee on Interior and Insular Affairs. Serial No. 93-40 (92-75). Washington, D.C.: Government Printing Office.

Interagency Review Group (IRG). 1979. *Report to the President by the Interagency Review Group on Nuclear Waste Management.* TID-29442. Washington, D.C.: Department of Energy, March.

Kellogg, W.W. 1975. "Climate Change and Influence of Man's Activities on the Global Climate." In *Energy and Man: Technical and Social Aspects of Man.* New York: Institute of Electrical and Electronics Engineers Press.

Kreitler, C.W., and T.C. Gustavson. 1976. "Geothermal Resources of the Texas Gulf Coast—Environmental Concerns Arising from the Production and Disposal of Geothermal Waters." Vol. 5. Proceedings of the Second Geopressurized Geothermal Energy Conference, University of Texas–Austin, February 23-25.

Lipschutz, R.D. 1980. *Radioactive Waste: Politics, Technology, and Risk.* Cambridge, Mass.: Ballinger Publishing Company.

Lovins, A.B. 1979. "Thorium Cycles and Proliferation." *Bulletin of the Atomic Scientists* 35, no. 2 (February).

MacDonald, G.J. 1979. "An Overview of the Impact of Carbon Dioxide on Climate." *Bulletin of the American Physical Society* 24, no. 1 (January):31.

Mankin, C.S. 1979. "Gas Resources and Reserves." *Environment* 21, no. 1. (January-February).

Marshall, E. 1979. "OPEC Prices Make Heavy Oil Look Profitable." *Science* 204 (June 22):1283-87.

Maugh, T.H. 1977. "Oil Shale: Prospects on the Upswing . . . Again." *Science* 198 (December 9).

Mechanical Engineering. 1976. "Geothermal Power May Trigger Earthquakes." February.

Metz, W.D. 1976a. "Fusion Research (I): What is the Program Buying the Country?" *Science* 192 (June 25):1320.

———. 1976b. "Fusion Research (II): Detailed Reactor Studies Identify More Problems." *Science* 193 (July 2):38.

———. 1976c. "Fusion Research (III): New Interest in Fusion-Assisted Breeders." *Science* 193 (July 23):307.

———. 1977. "Reprocessing: How Necessary Is It For the Near Future?" *Science* 196 (April 1).

———. 1978. "Mexico: The Premier Oil Discovery in the Western Hemisphere." *Science* 202 (December 22).

National Journal. 1979. "U.S. Mexican Relations May Hinge on Natural Gas Deal." II, no. 6 (February 10).

National Petroleum Council. 1972. "U.S. Energy Outlook, Interim Report: Initial Appraisal by the Oil Shale Task Group, 1971-1985." Washington, D.C.

New York Times. 1979. "Crude Oil Reserves Declined in 1978." April 30.

Nuclear Regulatory Commission (NRC). 1975. "Reactor Safety Study: An Assessment of Accident Risks in U.S. Commercial Nuclear Power Plants." WASH-1400. Washington, D.C., October.

Nucleonics Week. 1979. "Nuclear Costs Will Be Twice Those of Coal by 1986-87, says Komanoff." April 19.

Oil and Gas Journal. 1977. "U.S. Reserves Dip Blamed on Low, Uncertain Prices." April 18.

Parisi, A.J. 1979a. "Hope for Heavy Oil Producers." *New York Times,* July 19, p. D1.

———. 1979b. "Nuclear Power: The Bottom Line Gets Fuzzier." *New York Times,* April 8.

Post, R.F. 1976. "Nuclear Fusion." *Annual Review of Energy* 1: 213.

Rose, D.R., and M. Feirtag. 1976. "The Prospect For Fusion." *Technology Review,* December.

Smith, J.W., and H.B. Jensen. 1976. "Oil Shale." In *Encyclopedia of Energy,* p. 536. New York: McGraw-Hill Book Company.

Stickley, C.M. 1978. "Laser Fusion." *Physics Today* 31 (May):50.

Stobaugh. R., and D. Yergin. 1979. "After the Second Shock: Pragmatic Energy Strategies." *Foreign Affairs* 57, no. 4 (Spring):836.

Sullivan, W. 1979. "Fusion: The Answer to Fission." *New York Times*, May 15.

Tiratsoo, E.N. 1973. *Oilfields of the World.* Beaconsfield (England): Scientific Press, Ltd.

Union of Concerned Scientists (UCS). 1975. *The Nuclear Fuel Cycle.* Cambridge, Mass.: MIT Press.

——. 1977. *The Risks of Nuclear Power Reactors.* Cambridge, Mass.

United Nations (UN). 1974. *Survey of Energy Resources.* New York: U.S. Committee of the World Energy Conference.

U.S. Geological Survey (USGS). 1975a. "Geological Estimates of Undiscovered Recoverable Oil and Gas Resources in the United States." Circular 725. Reston, Va.

——. 1975b. "Assessment of Geothermal Resources of the United States—1975." Circular 726. Reston, Va.

Weinberg, A. 1976. "Can We Do Without Uranium?" Paper Presented at Oak Ridge Associated Universities Conference in Future Strategies for Energy Development, Oak Ridge, Tennessee, October 21.

White House. 1979. "Fact Sheet on the President's Import Reduction Program." Washington, D.C.: Office of the White House Press Secretary, July 16.

Willrich, M., and T.B. Taylor. 1974. *Nuclear Theft: Risks and Safeguards.* Cambridge, Mass.: Ballinger Publishing Company.

Wilson, C.L. 1977. *Energy: Global Prospects 1985-2000.* New York: McGraw-Hill Company.

Yonas, G. 1978. "Fusion Power With Particle Beams." *Scientific American.* 239, no. 5 (November).

Chapter 4

Renewable Energy Resources

Renewable energy resources are nondepletable and will continue to be available indefinitely. Although these energy resources are essentially inexhaustible in terms of their future availability, there are strict limits placed on the amount of energy available per unit of time. A small amount of renewable energy is supplied through the tides by the kinetic and gravitational energy of the earth-moon-sun system. By far the world's most important inexhaustible source, however, is solar energy, which can be harnessed both in the direct form of sunlight or in more indirect forms as the energy stored in the wind, plants, and water impounded in elevated reservoirs.

Solar energy drives the earth's climatic system and assumes many forms in addition to that of direct sunlight. For example, the kinetic energy contained in the winds is induced and sustained by the uneven heating of the earth's surface by the sun. Living plants convert sunlight to chemical energy through the process of photosynthesis. Hydroelectric energy, the kinetic energy of falling water, is made available on a renewable basis by the solar-powered evaporation of water. Ocean thermal gradients, another potential energy source, arise due to the solar heating of ocean surface water, coupled with the large-scale circulation of ocean waters driven by the sun. Other indirect sources of solar energy, including waves, ocean currents, and salinity gradients, are discussed later in this chapter.

The sun is by no means a "new" and "exotic" energy source. Solar energy, by supporting photosynthetic plant growth, provides for all the food we eat, thereby making life possible on the planet. Firewood—the major source of energy in the United States through the nineteenth century and still the principal energy source for much of the developing world—is an indirect form of solar energy. Even the energy contained in fossil fuels, comprising 90 percent of the current U.S. energy supply, was originally derived from the sun. In fact, solar energy is presently a major input to the economic systems of all countries, even highly industrialized ones (Taylor 1977). In the United States, the amount of solar energy used annually to grow plants, some of which are used for the production of food, lumber, paper, and other products, is approximately twice the energy obtained from "conventional" sources—oil, gas, coal, and nuclear energy. Therefore, recent proposals to harness solar energy do not involve tapping into a new energy source, but rather imply extending the range of applications for what already constitutes our most significant resource.

The virtues of solar energy are readily apparent: it is renewable, nonpolluting, abundant, and widely distributed. On an annual basis, approximately 44,000 quads of sunlight fall of the U.S. land mass; in comparison, total energy use in the United States was only about 79 quads in 1978. Thus the most important questions concerning solar energy utilization relate not to the magnitude of the solar resource, but rather to whether energy can practically and economically be made available in appropriate forms at the required time and locations. The distributed nature of the solar resource makes possible the collection of energy at or near point of end use, in some cases obviating the need for energy transmission, with its attendant costs and energy losses. Contrary to some misconceptions, significant quantities of solar energy are available across the country, not only in the sunny Southwest. For example, Portland, Maine, far in the northeast corner of the United States, receives on the average nearly 70 percent as much solar energy per year as does Tucson, Arizona, situated in the heart of the "sunbelt." There is, in fact, rarely more than a factor of two difference between the yearly amount of solar energy incident on an extremely favorable location in the United States as compared to that available at a highly unfavorable site.

Solar energy has some drawbacks as well. It is a dilute and variable source, and thus the difficulties involved in its utilization are far from trivial. The diffuse nature of solar energy means that significant amounts of land must be devoted to energy collection. Nonetheless, land use should not pose an insuperable barrier to the widespread utilization of solar energy. With 40 percent thermally

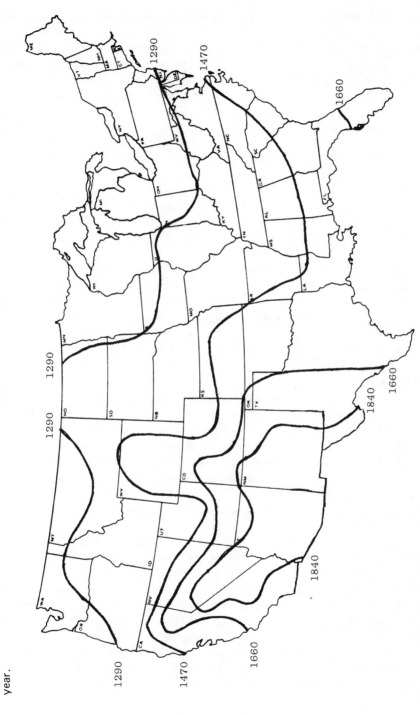

Map 4-1: Mean Daily Solar Radiation (Btu/square foot-day) Over the United States. The numbers are daily averages over a year.

Source: Solar Energy Research Institute.

efficient solar collectors, less than 0.5 percent of the total U.S. land area would need to be devoted to energy collection in order to provide all of the nation's current energy needs. This amount of land is less than half that presently devoted to roads and highways and solar utilization would not necessarily preclude other uses.

The intermittent nature of solar energy means that some method of storage will ultimately be essential to ensure the availability of energy when it is needed. Problems involved with integrating fluctuating sources of energy with conventional energy supply systems also need to be resolved before the widespread introduction of many solar technologies. The development of the numerous storage technologies that will ultimately be required remains a major technical challenge for a long-range solar economy. However, as discussed in Chapter 5, the fact that many storage systems are not yet commercially available should not significantly impede the near-term utilization of solar energy.

Although the solar resource is prodigious, there are limits to the amount of energy that can be extracted even from sources that are continually replenished. While renewable energy fluxes can often be harnessed with negligible environmental impact, major climatic shifts could ensue if exploitation were to proceed beyond a certain threshold level. Estimates of the energy potential for the major solar sources are presented in Table 4-1 and illustrated in Figure 4-1.

Solar technologies are not equally benign from an environmental standpoint. While solar energy can indeed be used with low environmental impacts in comparison to conventional energy sources, some modes of solar utilization—such as large-scale monocultural biomass plantations, extensive use of ocean thermal gradients, and solar power satellites beaming microwaves to earth—are markedly inferior to others in this regard.

With thoughtful planning and design, however, the operation of virtually all solar technologies can be made to occur with little or no generation of pollution or other waste by-products. The principal environmental effects will be incurred in the production of solar energy-related equipment. Because much less energy is normally required for the manufacture of these devices than will be produced over their useful lifetime, the amount of pollution incurred in manufacturing will be more than offset by the pollution-free operation of the solar devices. In the long term, even this impact could be greatly diminished, if not eliminated, by the establishment of a solar "breeder" economy, in which solar energy is utilized to produce solar (and other) equipment.

Table 4-1.   Renewable Energy Resource Estimates.

Source	Annual Renewable Energy (quads/year)
Direct Solar	20–200[a]
Wind	10–40[b]
Hydroelectric	3–4[c]
Biomass: Plantations	2–10[d]
Residues and Wastes	2–10[e]
Ocean Thermal Energy Conversion (OTEC)	0–20[f]
Tides	$\ll 1$[g]
Waves	≤ 1[h]
Ocean Currents	$\ll 1$[i]
Salinity Gradients	$\ll 1$[j]

[a]*Direct Solar*

Low Estimate—20 quads/year. Assumes that 0.1 percent of the U.S. land area is covered with 45 percent efficient (on a primary energy equivalent basis) solar collectors and an average solar insulation rate of 60 Btu/ft^2/hr.

High Estimate—200 quads/year. Assumes that 1 percent of the U.S. land area is covered with 45 percent efficient solar collectors. (This value does not represent a physical upper limit, but nonetheless should be more than sufficient to supply all our energy needs for the foreseeable future.)

[b]*Wind*

Low Estimate—10 quads/year. According to one study, this is the amount of primary energy that could be provided by wind generators located at sites selected on the basis of high average wind speeds on land not presently being used or planned for use (Merriam 1978).

High Estimate—40 quads/year. Estimates made for the extractable wind energy potential of the Great Plains (Blackwell and Feltz 1975; Eldridge 1975), the Atlantic coast (Heronemus 1976), and the state of California (Craig et al. 1978) alone amount to more than 40 quads. In addition, locations favorable for wind energy conversion also exist in the Pacific Northwest, the Great Lakes region, the Texas Gulf coast, and parts of Alaska.

While 40 quads per year may indeed represent a maximum practical limit, it does not approach the physical limit of the wind resource. For example, Gustavson (1979) has estimated that harnessing 10 percent of the wind energy naturally dissipated in the atmosphere within one kilometer of the U.S. land mass could annually provide 60 quads of electricity.

[c]*Hydroelectric*

Low Estimate—3 quads/year. Represents the current average value of hydroelectric energy production.

High Estimate—4 quads/year. Assumes that an additional 1 quad can be made available in an environmentally acceptable manner by upgrading existing hydroelectric facilities and installing generation equipment at existing dams presently used for other purposes. Contributions can also be made from the installation of small-scale, "low head" hydroelectric facilities; this resource has not been definitively assessed, but the total amount of energy that can be provided with minimal impact on the environment is generally considered to be small.

The 4 quads per year estimate is consistent with that made by the Department of Energy (1978).

[d]*Biomass Plantations*

Low Estimate—2 quads/year. Assumes that twenty-three million acres of land, or about 1 percent of the total U.S. land area, are devoted to the cultivation of energy crops, producing yields of fifteen dry tons of organic matter per acre per year (corresponding to a solar conversion efficiency of about 1 percent) that are converted to fuels with an average efficiency of 40 percent.

High Estimate—10 quads/year. Assumes that the same acreage is devoted to growing high yield energy crops capable of capturing solar energy with an average efficiency of 2 percent and producing thirty dry tons of organic matter per acre per year. In order to produce 10 quads of biofuels, the full energy potential of the biomass feedstock material would have to be realized by means of a conversion process such as that proposed by Antal (1976). Antal's technique would involve the use of solar heat to convert the organic matter into fuel form in a reaction called steam pyrolysis. The energy content of the fuel produced in this manner, augmented by the input of solar thermal energy, could exceed that of the original organic feedstock.

The amount of acreage devoted to growing energy crops might possibly be expanded beyond the twenty-three million acre limit arbitrarily assumed. However, this mode of solar energy utilization places severe demands on land and water resources, and before further expansions in biomass energy plantations are contemplated, land and water use considerations should be seriously confronted, especially in light of conflicting needs for these valuable resources. For example, one possibility for extensive biomass plantations is offered by the more than 100 million acres of land chronically afflicted with wetness problems (Poole and Williams 1976). The economic feasibility and environmental desirability of making use of these lands for the growing of energy crops should be thoroughly assessed.

[e]*Biomass Residues and Wastes*

Low Estimate—2 quads/year. Assuming that the total resource—consisting of agricultural crop residues, forestry and wood industry wastes, feed lot manure wastes, and urban refuse—amounts to about 11.5 quads (Poole and Williams 1976) and that 40 percent of the potential resource is collected and subsequently converted to a useful fuel form with an average efficiency of 45 percent. If 80 percent of all the organic residues and wastes are collected and converted to fuels with an average efficiency of 50 percent, nearly 5 quads of useful energy could be provided.

High Estimate—10 quads/year. This estimate assumes that the 90 percent of the total biomass residue and waste resource is collected and then converted to fuel with nearly 100 percent efficiency, perhaps through an approach similar to that suggested by Antal (1976). This upper limit must be regarded as highly tentative. The practical feasibility of recovering such a high fraction of available wastes is not yet known, nor is it clear whether exploitation could occur at this level without seriously undermining soil quality. Furthermore, the technology for economically converting organic matter to fuels with such high efficiency remains to be demonstrated.

This upper limit is similar to an estimate made by Commoner (1979) that 8 to 10 quads of alcohol and methane could be made available on an annual basis by integrating biofuel production with livestock production rather than by specifically utilizing agricultural residues and wastes. Under this arrangement, crops now grown to support livestock production would first be fermented to produce ethyl alcohol (ethanol), with the residues from the fermentation process used as livestock fodder. Livestock manure would be collected and converted to methane.

[f]*Ocean Thermal Energy Conversion (OTEC)*

Lower Limit—0 quads. The technological and economic feasibility of OTEC systems has yet to be firmly established. Therefore, it is possible that no energy will be produced from this source.

Upper Limit—20 quads. No authoritative estimate has been made of the OTEC resource available to the United States. The Gulf Stream region off the Florida coast appears to be the only suitable region near the continental United States. Von Hippel and Williams (1975) have estimated that about 20 quads of electrical energy (on a primary fuel equivalent basis) could be generated, allowing a $0.5°F$ temperature drop of the Gulf Stream. It is not yet known whether significant climatic effects would occur at this level of exploitation. In light of the presently inadequate understanding of climatic mechanisms, this estimate should be regarded as a maximum upper limit. It should be further added that the technical and economic feasibility of large-scale OTEC plants remains to be demonstrated, making this upper case estimate highly optimistic at present.

[g]*Tides*

Upper Limit—less than 1 quad. The worldwide potential of harnessable tidal energy is estimated to lie between 1 to 2 quads (primary energy equivalent); only about 3 percent of this total hydraulic energy potential is available in sites in or adjacent to the United States—the Passamaquoddy Bay on the Maine-Canada border and the Cook Inlet in Alaska (Merriam 1978).

[h]*Waves*

Upper Limit—1 quad. Would require wave generators capable of converting the kinetic and gravitational potential energy of waves to electricity with a 50 percent efficiency spanning a length of 300 miles along the West Coast of the United States, assuming an average power density of 75 MW/mile (Leishman and Scobie 1976).

The contribution from wave energy could be expanded by extending this line of wave convertors along the West Coast and by installing multiple rows of wave convertors, spaced adequately to allow the waves to regenerate themselves between successive rows. Additional wave energy could be extracted from the East Coast of the United States whose wave energy potential is only slightly less than that of the West Coast (Leishman and Scobie 1976). However, given the currently immature state of the art of wave energy conversion technology, estimated contributions in excess of 1 quad are premature at the present time.

[i]*Ocean Currents*

Upper Limit—less than 1 quad. At the present time, ocean currents cannot be regarded as a source that can be practically harnessed to produce significant quantities of energy.

[j]*Salinity Gradients*

Upper Limit—less than 1 quad. Significant quantities of energy cannot be obtained from this resource in an environmentally acceptable manner. Demands for large quantities of highly pure water pose the most formidable constraint. For example, in order to produce 1 quad per year, salinity gradient power plants operating with an efficiency of 50 percent would annually consume about 190 trillion gallons (Craig et al. 1978)—equal to more than 40 percent of the annual river runoff in the United States (Forest Service 1975). Clearly, salinity gradients cannot be relied upon to provide more than extremely minute amounts of power.

Figure 4-1. Renewable Energy Resource Estimates. The size of the bars reflects the uncertainty in current estimates. A discussion is provided in Table 4-1 and associated notes.

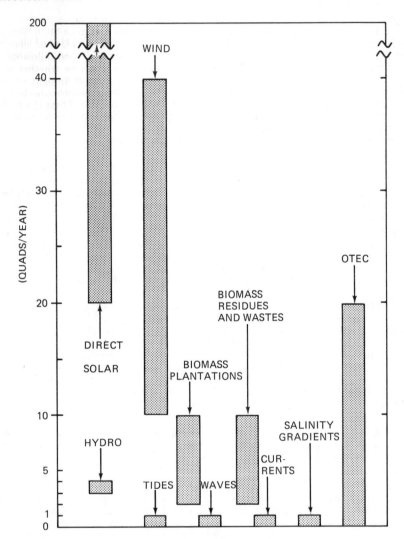

Material and resource constraints will be of some concern in a solar energy future, but are not expected to pose any insurmountable barriers to solar development. In those areas where specific materials are in critically short supply, the substitution of other, more abundant materials is generally possible. Resource constraints will probably force a reliance on a diverse mix of energy conversion and storage technologies and also encourage the recycling of critical materials—an activity that is favorable on environmental and energy efficiency grounds, regardless of resource limitations.

In contrast to all other energy sources, solar energy can be utilized without adding significant quantities of thermal energy to the earth. The installation of solar collectors can, however, lead to changes in the local heat balance. These modifications will be most significant for large, centralized solar-electric facilities and will be virtually undetectable for small, dispersed solar energy systems. However, the net heat burden imposed on the environment by the operation of solar energy devices will in any event be small in comparison to the thermal pollution caused by conventional energy systems of comparable magnitude.

Great flexibility exists for the utilization of solar energy. Systems can vary in scale from those designed to provide energy for a single home up to those supplying energy for a large factory or metropolitan area. Solar energy can be readily collected at low temperatures for building and water heating. Higher temperature thermal energy for industrial processes or electricity generation can be obtained by means of concentrating solar collectors. Sunlight can be directly converted to electricity in photovoltaic cells, and numerous other electrical generation options exist as well. Solar energy, embodied in the form of organic matter, can also be used to provide important liquid and gaseous fuels.

The numerous solar technologies that have been proposed are currently at varying stages of development. No major technical breakthroughs are required for their introduction, but considerable improvements are generally needed to enhance their economic appeal. While some solar technologies are economically attractive for use today or in the near future, the majority of systems are presently too expensive for general use. The relative cost competitiveness of solar technologies can be substantially improved if steps are taken to offset the market distortions caused by the high level of subsidies afforded to conventional energy forms. With the removal of the remaining barriers to solar utilization, production of solar devices can be scaled up significantly, providing a large potential for manufacturing cost reductions. Ultimately, the use of solar energy—a source whose "fuel" is free and stable in price—may provide substantial economic benefits by acting as a "hedge" against inflation, caused in large part by a reliance on depletable energy forms. We will first briefly review the prospects for the major approaches to solar utilization. More detailed assessments of the various solar resources and technologies are presented in subsequent sections.

Technologies designed to capture sunlight in the form of thermal energy can play a very important role, because thermal energy will constitute roughly half of our primary energy requirements in the future. Solar residential and commercial water and space heating systems are commercially available today and are already economically attractive in many parts of the country. Solar air-conditioning technology is somewhat less advanced but should become economically competitive in the near future and perhaps within a few years. The overall contribution in this area could be great, because energy requirements for residential and commercial water heating and space conditioning presently account for nearly 25 percent of total U.S. energy use. While most of the activity to date has concentrated on systems designed for individual buildings, community scale solar heating systems with centralized, annual storage hold great promise and are therefore deserving of much more attention.

The technology for providing low temperature solar thermal energy for agricultural and industrial process heat applications is available today, and the economics appear generally favorable. The potential contribution in this sector, however, is limited by the relatively low overall energy demand. Solar technologies capable of providing intermediate to high temperature thermal energy in the form of industrial process heat and steam can potentially make a much greater contribution, but the economic feasibility of this application awaits further progress. The fact that nearly 25 percent of all energy used in the United States is devoted to industrial process heat applications clearly justifies a greatly expanded, government-sponsored developmental effort.

A number of technologies designed to convert solar energy to electricity look attractive. Solar thermal energy can be converted to electricity by means of heat engines or by producing steam to drive conventional turbine generators. While this general concept appears promising, too much emphasis has been placed on large-scale, centralized power stations, and too little attention has been devoted to smaller scale, total energy systems that are located close to their loads and are capable of providing useful heat in addition to electricity. Photovoltaic or solar cells, which convert sunlight directly to electricity, offer numerous potential advantages. The most significant barrier to widespread utilization is the presently high costs of these devices. However, manufacturing cost reductions similar to those already achieved in related semiconductor industries could make solar cells economical for an expanding range of applications within a decade. Wind power systems represent the most technically and economically mature of the "new" solar-electric technologies. Wind power, already beginning to look economically competitive in certain parts of the country, has the potential to make a sizeable near-term contribution to the nation's electricity supply. Hydroelectric power generation currently represents by far the most important renewable means of electricity production. While hydropower systems will continue to make small, but important, contributions to the U.S. electricity supply, net production of hydroelectricity cannot be significantly expanded. The most important enlargement of hydropower's role will be as a means of providing large-scale energy storage capability.

Biomass, the solar energy stored in organic matter, represents another important solar resource. The major role for biomass in the future will be as a renewable source of liquid fuels for transportation. In fact, small amounts of alcohol produced by the fermentation of grains are already being used today to extend modestly supplies of gasoline. Organic matter can also be used to produce the gaseous fuel methane and to provide critical carbon feedstocks for petrochemical industries. Unfortunately, the biomass resource is constrained by fundamental limitations. The cultivation of crops strictly for energy purposes can directly compete, in terms of land and water use, with the production of food and lumber. The integration of biomass production with agricultural and forestry industries and solid waste management offers a more attractive approach, but the amount of biofuels that could be made available by this means is inherently limited. Although the ultimate quantity of fuels that can be produced from biomass sources is highly uncertain at the present time, it is nonetheless clear that the role played by biomass will be a critical one, regardless of the precise level of its net productive output.

The prospects for other proposed solar-electric technologies are much less favorable. Orbitting solar power satellites, beaming energy back to earth in the form of microwaves, presently look unattractive owing to their very high development costs and unanswered questions pertaining to the concept's economic feasibility and environmental acceptability. Similarly, the technological and economic feasibility of ocean thermal energy conversion (OTEC) plants—tapping the temperature gradients within the ocean as a power source—remains to be firmly established. Furthermore, the environmental consequences of large-scale energy extraction from ocean thermal gradients needs to be more fully explored prior to an extensive commitment to the OTEC concept. Tidal power generation is technologically proven, but offers very limited energy potential for the United States and the world in general.

Wave conversion technology is presently too immature to support any confident predictions regarding the ultimate potential of the wave resource. Even if commercially viable technology can be developed, wave power could make only a limited contribution to the U.S. energy supply, of importance on strictly a regional basis. Ocean currents do not appear to represent an energy source that can be harnessed to produce appreciable quantities of energy. Finally, salinity gradients can be ruled out as a significant energy source owing to the unacceptably large quantities of fresh water needed for conversion to electricity.

SOLAR THERMAL ENERGY

Introduction

The simplest method of harnessing solar energy is to collect and utilize it in the form of heat. The captured thermal energy can be used to perform a variety of tasks spanning a broad range of required temperatures. Relatively low temperature (less than about 212°F) tasks include domestic water heating, building heating, air conditioning, and agricultural crop drying. Intermediate (212-572°F) to high (exceeding 572°F) temperatures are required for most industrial process heat applications.[1] Low, intermediate, and high temperature applications are respectively, about one-half, one-sixth, and one-third of total thermal energy use. The utilization of solar energy for residential and commercial water heating, building heating and cooling, and industrial process heat applications can be of major significance, because taken together, these tasks account for nearly half of all energy currently used in the country.

Solar thermal energy systems can make an important contribution to the nation's energy needs in this century. Residential and commercial systems will be the first to assume a sizeable role. In 1977, roughly 65,000 solar hot water and/or space heating systems were sold in the United States; sales continued at this rate through 1978 and accelerated in 1979 (Maidique 1979). There are presently between 40,000 and 50,000 solar installations in the state of California alone (*Solarwork* 1979). The federal government has established a goal of having two and a half million solar homes by 1985, and California is hoping to have one and a half million solar-equipped homes and businesses by the late 1980s. Low temperature agricultural and industrial systems are fairly well advanced, but significant

1. The conversion of high temperature solar thermal energy into electricity is discussed in the next section.

penetration into this market has not yet occurred. Additional work is needed for the widespread introduction of intermediate and high temperature systems for industrial process heat and steam applications, but these systems may become ready for commercial use within a decade. Energy storage is an area where progress would be particularly important.

Technology

Solar water heating and building space heating systems presently represent the most technologically advanced and economically attractive solar energy applications. Solar heating and cooling systems can be placed into two general categories—passive and active.

Passive solar heating and cooling systems basically consist of highly energy-efficient buildings designed in such a way as to minimize heat losses in the winter and heat gains in the summer.[2] Passive systems utilize natural energy flows to transfer heat into and out of buildings, without a reliance on the forced circulation of a heating or cooling fluid. These systems often rely on large south-facing windows to increase the gain of solar energy. Insulated shutters mitigate window heat losses during winter nights, whereas reflectors, overhangs, and other shading devices diminish unwanted solar radiation during the summer. Thermal energy is frequently stored in the building floor, walls, and/or ceiling.

In some passive systems, solar collectors, separate from the basic building structure and powered solely by natural convection, provide warm air or water to the space to be heated. In other systems, solar energy is both collected and stored by darkly painted concrete walls or walls composed of blackened, water-bearing drums. Roof ponds can employ water-filled plastic bags installed on ceilings to store thermal energy and provide space heating and cooling. Passive solar energy utilization can also be accomplished by greenhouses designed to permit excess collected heat to migrate into the buildings to which they are attached. An innovative solar home built recently at the Massachusetts Institute of Technology stores thermal energy in ceiling tiles filled with phase-changing material (Dietz 1978). This passive system is expected to derive 85 percent of the building energy requirements from the sun.

Active, as opposed to passive, solar heating and cooling systems rely on an external power source for forced transport to distribute

---

2. Anderson and Michal (1978) provide a thorough discussion of the many types of passive solar heating and cooling systems.

energy within buildings. Special solar collectors and thermal storage reservoirs, apart from the normal heating systems, are employed. Active solar water heating and space heating systems are widely available on a commercial basis. Approximately 200 firms are currently selling these systems in the United States (CEQ 1978).

The flatplate collector is most commonly used in present systems. This device consists of a blackened metal absorber, mounted on an insulated supporting structure, with glass or plastic covers. The basic principle of operation is quite simple. Solar radiation passing through the transparent cover is absorbed by the metal plate, causing the plate to heat up. Heat from the collector is transferred to a circulating fluid, such as water or air. Temperatures ranging from 100°F to about 200°F are commonly achieved with such a device. The thermal energy contained in the heated fluid can be used directly for service water heating and space heating, or it may be stored for subsequent use in water, rock, or some other medium.

Photograph 4-1: Solar Thermal Heating and Cooling. This heating and cooling system uses solar collectors to provide 60 percent of the year-round requirements of the George A. Town Elementary School in Atlanta, Georgia.

Source: Westinghouse Electric.

While the flatplate collector represents a well-established, commercial technology, the concept leaves little room for dramatic cost reductions owing to the relatively fixed amounts of metal and other expensive materials used in their manufacture (Hammond and Metz 1978). Consequently, serious competition is posed by collectors using inexpensive plastic and rubber materials. At the other end of the spectrum, competition is presented by highly efficient, advanced collectors capable of providing higher temperature heat. Prominent among the latter category are evacuated tube collectors in which the absorbers are enclosed by a glass cylinder within a vacuum. These collectors are capable of achieving temperatures in excess of 200°F with an operating efficiency of 50 percent.

Community scale solar heating systems with centralized annual thermal energy storage hold considerable promise. Centralized storage is attractive owing to its reduced heat losses and costs per unit of stored energy. This is because larger systems have higher volume-to-area ratios than small systems, whereas both storage costs and heat losses are roughly proportional to the surface area of the storage vessel. One scheme for providing space heat for 800 houses has been proposed in West Germany (*Audubon* 1975). This particular system would employ solar energy to heat an insulated pond (275 feet in diameter and 30 feet deep) to a temperature of up to 170°F. A solar pond system built in Miamisburg, Ohio, in August 1978 is being operated experimentally for the Department of Energy. The pond is used to deliver 500,000 Btu's per hour to the city's swimming pool (SEIR 1979a). District heating systems employing ponds for the collection and storage of solar energy have also been suggested by Taylor (1978). In one proposed system, hot water is collected by a thirty-foot deep pond; insulation is provided by the ground, and heat losses are mitigated by a transparent, floating plastic cover. A community energy system suggested by Engelke (1978) would serve 500 to 10,000 homes through the interseasonal storage of solar thermal energy in hot water reservoirs. Work is also underway in Sweden to develop district heating systems storing solar-heated water in underground, insulated pits (Margen 1978). A small-scale demonstration plant was commissioned in 1978, and a fifty home heating system is scheduled to be built in 1979.

A heat pump can be used to back up a solar heating system (Glicksman 1978). A combined solar–heat pump system has several potential advantages. By using solar-heated water as the low temperature heat source, the heat pump could operate at a high efficiency. By operating at relatively low temperature, the solar collector system could also achieve a high average efficiency.

Active solar cooling is less technologically advanced than solar heating and needs additional development to become economically competitive. Solar-powered air conditioning is, nonetheless, technically feasible, and one system is now commercially available from Arkla Industries (SED 1977). Active solar cooling systems are more complex than heating systems, requiring temperatures exceeding 180°F for operation. Of the various systems under consideration, absorption cooling systems using lithium bromide-water or ammonia-water as the working fluid are considered to be the most promising. One attribute of solar cooling is the close correlation between high solar insolation and high cooling demand, meaning that sunshine is generally abundant on days when cooling is most needed. A much simpler method of providing air conditioning or refrigeration in regions with a cold climate would be to store ice produced in the winter and use it for cooling purposes year round (Taylor 1978).

Solar thermal energy can also provide process heat for agricultural and industrial applications requiring hot water, air, and steam over a wide range of temperatures. Agricultural process heat systems for applications such as grain drying normally operate at temperatures below 212°F (DOE 1978), permitting the use of flatplate collector systems in many cases.

Temperature requirements for industrial process applications span a broad, but generally higher, range. Approximately 30 percent of all industrial process heat is used at temperatures below 572°F (Metz 1976). One study estimated that by the year 2000, 7.5 quads of solar energy could be used for this application (ITC 1977). Temperatures approaching this value can be achieved with stationary, concentrating solar collectors, such as the compound parabolic collector (Winston 1976). Somewhat higher temperatures can be attained with one axis tracking concentrators. More than a half dozen firms are currently manufacturing collectors in the latter category (Hammond and Metz 1978). Several demonstration projects of industrial process applications have been initiated by ERDA and DOE (DOE 1978). One of the projects will employ parabolic trough collectors to provide 185°F water for a Campbell Soup plant in Sacramento, California (Metz 1976).

Economics

Passive solar heating and cooling techniques are often highly cost-effective, in many cases adding little to the cost of a new building. The implementation of these techniques, however, is often not economically feasible for existing buildings. Prices for flatplate collectors vary considerably, but generally fall within a range of $5 to $15 per square foot (Herman and Cannon 1977). Plastic collectors, capable of roughly comparable performance, have the potential to be produced at much lower costs, some estimated at less than 50 cents per square foot (Taylor 1978). The price of evacuated tube collectors and intermediate temperature, single axis tracking concentrator systems currently sell for about $9 to $18 per square foot (Hammond and Metz 1978).

Economic analyses by the MITRE Corporation (1976), the Massachusetts Energy Policy Office (1976), Nadis (1977), and the Office of Technology Assessment (OTA 1978) all indicate that solar water and space heating are presently competitive with electric heating in many parts of the country and may be competitive with oil and gas heating in the 1980s. According to the OTA (1978) analysis, solar heating systems for both high-rise buildings and community scale systems (for 300 homes), connected to a central, seasonal thermal storage facility, have the potential to provide 100 percent of the space heating and hot water needs at costs competitive with conventional electric systems. The economics of district heating systems employing solar ponds also look quite attractive (Taylor 1978; Donovan et al. 1979). The cost of heat from the solar pond in Miamisburg, Ohio, is estimated to be equivalent to heating oil at a price of 65 cents a gallon (SEIR 1979). Heating oil is expected to cost 90 cents per gallon in the winter of 1979–1980.

Active solar air conditioning is presently more expensive than solar heating. However, solar cooling technology is relatively new, and consequently, significant cost reductions could occur with additional development (DOE 1978). By allowing for the utilization of solar collectors throughout the year, combined solar heating and cooling offers potentially significant economic advantages. Active solar air conditioning is likely to become competitive first in southern regions having long cooling seasons.

Solar agricultural process heat systems have relatively low costs at present and are likely to be economically competitive in the near future (DOE 1978). Solar industrial process heat systems are presently more expensive. Costs reductions from the mass production of solar equipment and rising fuel costs, however, could make these systems economically attractive in the near term.

Environmental Impacts

Residential, commercial, and industrial utilization of solar thermal energy will have minimal environmental impacts. Passive and active solar heating systems are, in fact, considered to be among the most benign of all energy technologies (Harte and Jassby 1978). Land use will not constitute a major problem, because in the majority of cases, the solar collectors can be integrated into the buildings using the collected energy. The operation of these systems will directly produce no pollution.

The principal environmental effects will be those associated with the manufacture of solar equipment. However, it is expected that throughout their useful lifetime, solar heating systems should be able to produce about twenty-five times more energy than was consumed in their production (SRI 1977). Therefore, the operation of nonpolluting solar thermal systems can lead to a significant net reduction in pollution, even if conventional energy is used for their manufacture. Furthermore, after the expiration of useful system life, the basic materials can be readily recycled to produce new equipment, requiring much less energy and producing correspondingly less pollution than manufacturing them from raw materials.

Substantial quantities of steel, glass, and aluminum will be required for the large-scale manufacture of solar collectors and related equipment. These materials generally appear to be in adequate supply to support large-scale deployment, although modest expansion of some mineral industries might be required (SRI 1977). If shortages in these materials arise, alternative materials are available (FEA 1974). In the long term, growth in primary resource demands can and should be curbed by material recycling.

Overall Assessment

Solar residential and commercial water and space heating systems are commercially available today and are already economically attractive in many parts of the country. Solar air-conditioning technology is currently less advanced, but could become economically competitive in the near future with additional development. Residential and commercial water heating and building space conditioning presently account for nearly 25 percent of total U.S. energy use, and consequently, the ultimate contribution from solar energy in meeting these needs is considerable. Although most development activity to date has concentrated on systems for individual buildings, it remains to be established whether, in fact, this is the optimal approach. Community scale solar heating systems with central, annual storage offer numerous potential advantages that should be carefully assessed prior to the widespread introduction of solar heating technologies.

The economics of low temperature solar thermal energy systems for the supply of agricultural process heat appear to be quite favorable. The ultimate contribution from solar energy in this application, however, will be limited by the relatively low overall energy demand in this sector.

Solar energy is well suited to providing the intermediate to high temperature thermal energy required for industrial process heat applications, but the economic feasibility of this application awaits further development. The slow progress achieved to date in this area is largely a result of limited support from the federal government. The relatively lethargic solar industrial development and utilization program is difficult to explain in light of the fact that nearly 25 percent of total energy use in the United States is for industrial process heat applications. Given the sizeable contribution that could be made from solar energy utilization in this sector, a vigorous, government-sponsored developmental effort should be promptly initiated.

SOLAR THERMAL-ELECTRIC CONVERSION

Introduction
Solar thermal-electric systems convert heat to electricity by means of conventional steam turbine generators or heat engines. In addition to providing electricity, solar-powered heat engines can also provide mechanical energy for applications such as pumping water for irrigation. Solar thermal conversion systems can be built on a variety of scales ranging from small-scale residential installations to large, centralized power stations. While most of the emphasis to date has been placed on large-scale designs, smaller systems, perhaps built on a community scale, hold considerable promise. A major advantage held by smaller systems located near their loads is that utilization of the heat left over from electric conversion is greatly facilitated. In fact, "total energy" systems— capable of providing both useful thermal energy and electricity for residential, commercial, and industrial use—constitute one of the most promising approaches to solar energy utilization.

Technology
High temperature solar thermal energy can be converted to electricity in a process similar to that employed by conventional power plants, except for the difference in the initial thermal energy source. Although a number of different approaches to solar thermal conversion exist, the current research effort has been dominated by the "power tower" concept (Caputo 1977; Metz 1977a; Smith 1976; Vant-Hull

and Hildebrandt 1976). In this system, a large array of steerable mirrors called heliostats track the sun and reflect sunlight onto a central receiver-boiler set on top of a tall tower. The absorbed solar energy is used to produce steam, which in turn is used to generate electricity in a conventional turbine generator. As presently envisioned these plants would be large, although the optimum size is still unknown. A 100 MWe facility would require as many as 10,000 heliostats covering more than one square mile (Caputo 1977; Metz 1977a). It is normally assumed that they will be sited in southwestern regions of the United States.

The heliostats currently under consideration will track the sun along two axes and concentrate sunlight by a factor of about 1,000, requiring a high degree of precision. Temperatures of about 900°F can be attained (CEQ 1978). A combined solar collection and electrical generation efficiency of about 15 percent is expected. Major reductions in the cost of heliostats will be necessary to make the presently expensive power towers competitive, since heliostats comprise approximately 60 percent of the total system cost (Metz 1977a). The design of a boiler capable of withstanding rapid variations in temperature and power density is also considered to pose a formidable engineering challenge.

In recent years, solar thermal conversion has received 40 percent of the total government funding for solar and wind electrical technologies, 60 to 70 percent of which has been devoted to the power tower (DOE 1978a; Metz 1977a). A 5 MWt test facility has recently been completed in Alberquerque, New Mexico, at a cost of about $21 million (Metz 1977a). A 10 MWe central receiver (power tower) pilot plant, estimated to cost $120 to $130 million, is expected to begin operation in Barstow, California, in 1981. A 100 MWe demonstration plant, costing a projected $350 to $400 million, should be completed by the mid-1980s (Herman and Cannon 1977).

Although present designs are based almost exclusively on large systems, a small (50 kWe) and relatively inexpensive power tower has been built and operated in Italy since 1965 (Williams 1974; Metz 1977a). A similar, 400 kWe system was scheduled to begin operation at the Georgia Institute of Technology in late 1977.

The other major approach to large-scale solar thermal conversion under consideration is called a distributed receiver system (AIAA 1975; Meinel and Meinel 1976; Williams 1974). In this configuration, concentrating collectors focus sunlight onto an absorber pipe, where a working fluid is heated and then pumped to a central turbine generator or heat engine. The distributed concept is inherently more flexible in scale than power towers. The advantages of smaller

Photograph 4-2: High-Temperature Solar Thermal Energy Conversion. Rear view of some of the 72 heliostat arrays at the Department of Energy's "Power Tower" Solar Thermal Test Facility located at Sandia Laboratories in Albuquerque, New Mexico. The heliostats track the sun and reflect sunlight onto a boiler in the tower, producing steam to generate electricity.

Source: Sandia Laboratories.

systems, in terms of reduced heat losses and piping costs, may override any economy of scale advantages held by very large systems.

A number of small heat engines could be used for this application including conventional Rankine cycle engines and the more advanced Brayton and Stirling designs (OTA 1978). While European firms are presently manufacturing the most advanced engines, interest has recently been picking up in the United States, and a number of U.S. companies are currently marketing these devices (CEQ 1978).

Among the more promising approaches to utilizing solar thermal energy are community and factory scale power plants capable of providing both usable heat and electricity. These "total energy" systems are likely to be superior to independent solar-electric and solar thermal energy systems on energy efficiency, economic, and environmental grounds. Among the proposed systems are solar energy-collecting ponds, which can produce electricity in addition to heat, provided that a fluid with a low boiling temperature, such as freon or ammonia, is used (Taylor 1978). Electric conversion efficiencies on the order of about 10 percent are possible for freon engines operating on temperature differentials typical for these ponds. Overall system efficiencies, however, would be much higher owing to the fact that "waste heat" from the engine's condenser would be discharged back to the pond. Electrical generation efficiencies can be improved, if necessary, through the use of concentrating solar collectors, which would increase the temperature gradients within the ponds.

Engelke (1978) has also proposed a community scale total energy system capable of serving 500 to 10,000 homes. Solar thermal energy would be collected and stored on a seasonal basis in a large-scale hot water reservoir. Air-conditioning and refrigeration needs would be met by the storage and distribution of icy water. Electricity would be generated by employing efficient, ammonia cycle heat engines to operate between the (200°F) hot water reservoir and the (32°F) ice water reservoir. By using freezing water as the low temperature "sink," reasonable generation efficiencies could be achieved without employing tracking and concentrating solar collectors. The proposed system would also rely on wind generators to provide supplemental electricity.

A total energy test facility is now operational at the Sandia Laboratory in New Mexico. The Department of Energy is sponsoring two $10 million total energy demonstration projects, currently under construction in Fort Hood, Texas, and Shenandoah, Georgia (DOE 1978b; Metz 1977b). Operation is expected to begin in 1980; each plant will produce 200 kW of electricity and 1.5 MW of thermal power. The Shenandoah total energy system will provide hot water, space heating and cooling, process steam, and electricity for a textile factory. A 1 MWe, community scale total energy system is scheduled for operation in the 1980s (CEQ 1978).

Solar-powered heat engines operating in intermediate temperature ranges (of about 300°F) can also be used to pump water for irrigation. The potential contribution is significant, because over 300,000 irrigation pumps are now used in the western United States, at an annual energy cost exceeding $700 million (Metz 1977b). The largest solar pumping facility presently in the United States is a 30 kWe project built by a private research company in Gila Bend, Arizona, in 1975. Although this system is not quite economically competitive, the potential for cost reductions appears great. It has been estimated that even with only limited production, the entire system cost could be reduced fourfold (Metz 1977b).

Economics

Considerable uncertainty exists over the eventual costs of solar thermal conversion systems. Preliminary estimates of the capital costs of solar thermal power plants range from about $1,200 to $2,800 per peak kWe[3] (Caputo 1977; Herman and Cannon 1977; Metz 1977a). Electrical generation cost estimates for central receiver plants would be competitive with conventional power plants by the mid-1990s. However, additional research and development on the design of low cost heliostats and high performance receiver materials will be needed in order to assess the accuracy of these cost projections. Progress is also needed for the development of economical high temperature thermal storage systems, but these systems are expected to be commercially available by 1985 (EPRI 1976).

The optimum size of centralized solar thermal-electric facilities remains unclear at present. Available evidence suggests that power towers will be more economical on a large scale, whereas distributed systems will be less expensive on a smaller scale, but the exact level at which one approach is favored over another is presently unknown.

Although the development of small-scale solar thermal power systems and total energy systems has suffered from relative neglect, there exists no conclusive evidence demonstrating a decisive cost advantage for large-scale systems. In fact, the relative economics of small-scale systems could substantially improve as a result of cost reductions from mass production of small heat engines. Although heat engines presently cost an average of about $1,000 per kWe, large-scale production could possibly reduce these costs to as low as $20 to $40 per kWe, comparable to that of a typical automobile engine (Caputo 1977; Metz 1977b; OTA 1978). One study indicated that cost reductions in small heat engines could make solar electricity competitive with conventional electricity in the mid-1980s (OTA 1978).

The OTA study also concluded that under certain favorable conditions, community scale total energy systems may be able to compete with conventional utility systems by the year 2000. More optimistically, Engelke (1978) concluded that a community scale total energy system using commercially available technology could provide energy at a cost below the current cost of fossil fuels. Another analysis also found the economics of a district energy system using solar ponds to supply both heat and electricity to be quite favorable (Donovan et al. 1979).

---

3. "Peak kilowatt" costs refer to the capital costs of the generating system per kilowatt of power produced under conditions of maximum solar insolation.

Environmental Impacts

The environmental impacts associated with solar thermal conversion need not be severe. The large amounts of land needed for large-scale, solar-electric generating plants would be the most serious. However, the land required by a solar power plant would be roughly comparable to that of a coal-fired plant with equivalent output, when the amount of land disturbed for coal mining through the useful life of the plant is taken into account (Von Hippel and Williams 1975; Caputo 1977). Although centralized solar power plants could have a disruptive impact on the local environment, the land impact would be considerably less destructive than that resulting from the surface mining of coal (Herman and Cannon 1977). Land use impacts would be less severe for small-scale systems, partly owing to the fact that, in many cases, these systems could be directly incorporated into buildings and other structures.

Large-scale solar thermal plants could affect the local thermal energy balance by increasing the proportion of solar radiation absorbed to that reflected on the earth's surface. Possible effects on the weather are unknown at the present time. If this problem proved to be significant, however, it would be theoretically possible to offset increases in global thermal energy input by adding reflective surfaces to areas around the collectors. In any event, any incremental addition of thermal energy to the global system from the operation of a solar thermal power plant would be small in comparison to the effect of conventional fossil fuel and nuclear plants (Caputo 1977). Any thermal effects from small-scale systems would be negligible in comparison to the impacts already resulting from road construction and urban, industrial, and agricultural development.

Like conventional power plants, solar thermal-electric plants using a steam cycle and wet cooling systems would consume large quantities of water. Water is in scarce supply in the Southwest, where large-scale solar plants are often contemplated, and cannot be made available in sufficient quantities for this application. Therefore, dry cooling towers, rejecting waste heat directly to the air, will probably be necessary. Power plants with dry cooling towers are, however, about 10 percent more costly and also 10 percent less efficient than plants employing wet cooling towers (Caputo 1977). Water demands could also be lessened by a reliance on Brayton cycle gas turbine generators (Harte and Jassby 1978).

Solar thermal power plants are nonpolluting. Although some pollution would be produced in the manufacture of the solar plant equipment, the amount emitted would be less by a factor of ten or more than that resulting from the operation of a comparable conventional power plant (Caputo 1977; Harte and Jassby 1978). This is largely attributable to the fact that over its entire useful life, a solar power plant would generate ten to twenty times more energy than was consumed in its production. Corresponding net energy ratios for nuclear power plants under a variety of assumptions range from about five to twenty (Rutty et al. 1975).

Overall Assessment

Although solar thermal conversion can potentially assume a vital role in the nation's energy system of the future, the precise nature of that role cannot be accurately predicted at present. The relative advantages and disadvantages of power towers versus more flexible, smaller scale systems need to be determined. The ultimate contribution from solar thermal power will also depend on its relative attractiveness compared to photovoltaic and wind-electric systems. Additional research and development is required to resolve these questions. While the costs of power tower demonstration plants currently under development are clearly too high for commercial feasibility, the potential for cost reductions by the mass production of heliostats, heat engines, and other equipment is considerable enough to give hope.

In light of the existing state of knowledge, the federal government's overwhelming emphasis on the large-scale power tower concept and relative inattention to other concepts is without satisfactory justification. Holding particular promise are total energy systems, which may ultimately provide the most satisfactory method of supplying electricity and thermal energy to residential, commercial, and industrial users. Consequently, this and other imaginative concepts should be fully explored before their development is precluded by massive commitments of capital, materials, and human effort to the power tower concept.

PHOTOVOLTAIC CELLS

Introduction

Photovoltaic cells, commonly called solar cells, are semiconductor devices that can convert sunlight directly into electricity. These devices are attractive for a number of reasons. They have no moving parts and therefore are quiet and reliable, require little maintenance, and have the potential for long useful lives. Electrical generation produces no pollution and consumes no water.

Great flexibility exists for the utilization of photovoltaic devices. The cells are inherently modular, and consequently, systems can be designed on a variety of scales ranging from small-scale residential to highly centralized, large-scale facilities. While the production of solar cells is subject to economies of scale, the utilization of these devices is not (Kelly 1978). In fact, on-site photovoltaic systems hold numerous advantages over remotely sited, centralized plants. Systems can be installed directly on building roofs, thereby reducing land use requirements. Siting near the point of end use can in some cases reduce the need for electrical transmission and distribution equipment and also facilitate the utilization of thermal energy in photovoltaic total energy systems.

The technical feasibility of photovoltaic generation has been amply demonstrated. The first modern solar cell was developed at Bell Laboratories in 1954 (Herman and Cannon 1977). Since the late 1950s, these devices have powered the majority of U.S. space satellites. They have also been used for limited terrestrial applications to provide electricity at remote installations where other power supplies are not available. Over recent years, U.S. solar cell manufacturers have produced only about 500 kWe of generating capacity annually (Herman and Cannon 1977).

The ultimate potential for photovoltaic generation of electricity is truly vast. There are a number of promising options, including a variety of cell materials and array configurations, and it is too early to predict which approaches will ultimately win out. The principal barrier to extensive use is the present high cost of these devices. However, great potential exists for technical improvements and cost reductions in cell manufacture, leaving the outlook for photovoltaics quite promising. As with other solar-electric technologies, problems relating to energy storage and integration with conventional utility grid systems also need to be resolved before widespread implementation can occur, but this barrier should not severely impede near-term usage.

Technology

The theoretical principles of photovoltaic conversion are reasonably well understood at the present time and have been described in considerable detail by Chalmers (1976), Fan (1978), and Merrigan (1975). If light photons of sufficient energy are absorbed by a semiconducting material, electrons can be dislodged from their fixed positions within the lattice structure, leaving behind positively charged "holes." These promoted electrons have sufficient energy to move freely within the material. In order to derive useful electricity from this

Photograph 4-3: Photovoltaic Cells. An array of 120,000 solar cells provides 25 kilowatts of electric power to drive a 10 horsepower pump at an experimental irrigation project near Mead, Nebraska. The system irrigates about 80 acres of cropland.

Source: Department of Energy.

electron mobility, the cell is designed to have a built-in voltage that prevents the majority of electrons from recombining with holes until they have traveled through an external circuit connected to the load.

This internal voltage can be established by creating a junction where an abrupt change of conductivity occurs. In cells composed of a single material, such as silicon, a so-called "homojunction" is formed by adding different impurities or "dopants" to the pure material on opposite sides of the junction. A "heterojunction" can be formed by joining together two different semiconductors such as cadmium sulfide and copper sulfide. A "schottky junction" is formed by joining a semiconductor and a metal, such as amorphous silicon and platinum.

The theoretical efficiency of a solar cell is limited by the fact that solar photons lacking sufficient energy to release an electron from its bound state cannot contribute to the generation of electricity. Furthermore, photon energy exceeding that required to free electrons for conduction is converted into heat, rather than electricity. Any real device will, of course, fall short of the theoretical maximum for a number of reasons, including the reflection of light at the cell's surface and the fact that a small fraction of the surface will be covered with an opaque electrical contact (Chalmers 1976). The maximum theoretical efficiency for silicon cells is about 22 percent (Fan 1978); real devices can achieve values up to about 18 percent (Chalmers 1976). Upper limit conversion efficiencies for gallium arsenide and cadmium sulfide – copper sulfide cells are estimated to be about 26 percent (Fan 1978) and 15 percent (Kelly 1978), respectively.

The overwhelming majority of solar cells currently in use or being sold are single crystal silicon devices. Presently available cells are quite expensive, but much can be done to trim costs. For example, much of the work that goes into the fabrication of commercial cells is now done by hand. The automation of cell manufacture and assembly could lead to substantial cost reductions (Hammond 1977). The high cost of present cells is also partly attributable to the fact that the quality of semiconductor grade silicon now used in cell manufacture is much higher than is necessary (Chalmers 1976; Fan 1978). Lower quality silicon crystals would cost much less, with only a modest sacrifice in cell efficiency.

Another important cost factor stems from current production techniques that are very inefficient with respect to the use of energy and materials. The conventional practice is to grow cylindrical ingots of silicon, which are then sliced into thin wafers. More than 60 percent of the original silicon material, produced with a considerable expenditure of energy, is typically wasted in this process (Fan 1978). A much more efficient approach would be to grow continuous, crystaline sheets from molten silicon. Several methods exist for the growing of continuous silicon ribbons, including the edge-defined film-fed growth (EFG) technique under development at Mobil-Tyco Laboratories (Mlavsky 1976) and Westinghouse Electric's web-dendritic growth process. Cells prepared from these processes have achieved conversion efficiencies of over 12 percent (Fan 1978) and 15 percent (SEIR 1978c), respectively.

Single crystal silicon absorbs light poorly, and consequently, cells made out of this material have to be relatively thick—about 100 to 200 micrometers or about four- to eight-thousanths of an inch (Kelly 1978). The use of polycrystalline material can eliminate the costly process of growing a single, perfect crystal. Polycrystalline silicon can be formed by a variety of relatively inexpensive processes. One of the proposed methods would involve dipping a solid metal backing (sub-strate) into molten silicon, leaving a thin film of polycrystalline silicon on the substrate (Fan 1978). These "thin film" cells need be only 20 to 30 micrometers thick and, therefore, would require less silicon overall. Solar cells with efficiencies as high as 9.5 percent have been produced in this manner (Fan 1978).

Cells composed of a noncrystalline, "amorphous" silicon and hydrogen alloy represent a highly promising development. Conversion efficiencies of 6 percent have been reported for devices of this type (Hammond 1977), and a value of 15 percent is considered possible (Kelly 1978). This material strongly absorbs light, and typical cells require a thickness of only one micrometer, or only about 1 percent that required by conventional solar cells (Kelly 1978). Therefore, the advent of amorphous silicon cells could dramatically reduce overall silicon requirements. Material and manufacturing costs for these cells could be correspondingly low (Fan 1978).

Relatively thin solar cells could also be made out of other materials, such as cadmium sulfide and gallium arsenide, which absorb sunlight much more effectively than silicon. Major savings in both materials and manufacturing costs can be achieved by producing solar cells with "thin film" techniques. Although a number of options exist, the most common method involves the spraying or vapor deposition of a thin coating of a polycrystalline semiconductor material onto a metal substrate.

The only thin film cells commercially available at present use cadmium sulfide. The major drawback of these cells is low conversion efficiencies, necessitating larger solar cell array areas than would be required by silicon cells capable of producing the same power.

The efficiency of commercial cells is less than 5 percent (Hammond 1977). Laboratory cells, however, have achieved efficiencies of 8.5 percent, and a value of 10 percent is considered possible (SEIR 1978a). Even if low cost, thin film cells are produced, inexpensive supportive structures and encapsulation methods will be required to hold down overall systems costs. Reductions in noncell area costs will be an essential prerequisite to the widespread utilization of low efficiency solar cells.

Gallium arsenide appears to be a promising material for highly efficient, single crystal and thin film solar cells. Cells constructed out of this material could be much thinner (only about 1 to 2 micrometers thick) and lighter than comparable silicon cells (Fan 1978). Single crystal gallium arsenide cells have achieved conversion efficiencies of 22 percent. Thin film cells have so far been only able to achieve a 5.5 percent efficiency, although a figure of 15 percent is considered attainable (Fan 1978).

The use of concentrating solar collectors in conjunction with solar cells represents an alternate approach to reducing overall system costs. The collectors employed can range from nontracking systems capable of concentrating sunlight by a factor of only ten up to two axis tracking systems, using parabolic mirrors or fresnel lenses, with concentration ratios up to 1,000 or more. Because the solar-concentrating equipment will tend to dominate the total system cost, the use of highly efficient solar cells will be important in order to minimize the required collector area. Cells designed for operation at high solar concentrations differ considerably from ordinary cells. Optimally designed silicon cells operated in sunlight concentrated by a factor of 300 have achieved efficiencies as high as 18 percent (Kelly 1978). Gallium arsenide cells have achieved efficiencies of 24.6 percent under sunlight concentrated 180-fold (Fan 1978).

As cell temperatures rise above a certain critical point, conversion efficiencies fall off almost linearly. For this reason, active cell cooling will be required in many concentrating systems. This opens up the attractive possibility of utilizing the collected low temperature heat, in addition to the electricity generated, in so-called photovoltaic total energy systems. This approach looks very promising for those applications where a use for low temperature thermal energy exists (Kelly 1978).

Because solar cells cover only a small fraction of the total collector area in concentrating systems, more can be spent per unit area on individual cells, allowing for greater flexibility and innovation in the design of high performance cells. One imaginative approach would be to stack two or more different types of cells on top of each other, in

Photograph 4-4: Photovoltaic Concentrator. The plastic fresnel lenses at right are mounted so as to concentrate sunlight on the silicon solar cells at left. Each lens concentrates the equivalent of 50 suns on the corresponding cell. This array is located at Sandia Laboratories.

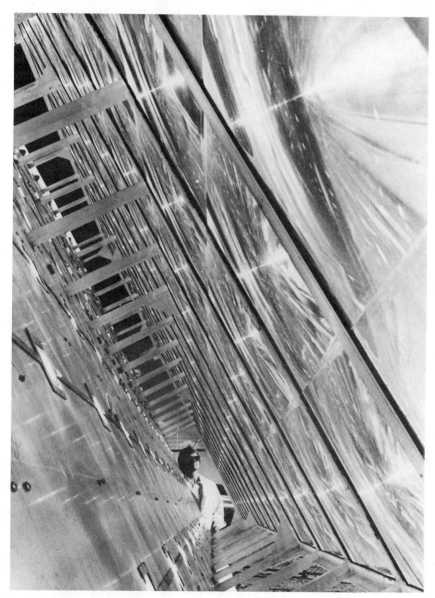

Source: Sandia Laboratories.

order to make far greater use of the solar spectrum than would be possible for an individual cell. A conversion efficiency of 28.5 percent has been achieved by stacking gallium arsenide and silicon cells (SED 1978a). The maximum attainable efficiency for this multi-cell approach is considered to be about 40 percent (Hammond 1977). Rather than vertically stacking cells, another option would be to employ optical filters that split the solar spectrum into different frequency bands that are directed to respective solar cells optimally designed for light of that particular frequency. One system of this type has achieved a 28.5 percent conversion efficiency; 40 percent is considered to be a practical upper limit (Fan 1978). A similar proposal calls for the use of fluorescent dyes capable of absorbing sunlight and reradiating light in several narrow frequency bands, each suitable for the efficient operation of a particular solar cell (*New Scientist* 1977). A maximum conversion efficiency of about 32 percent has been estimated for this concept.

The somewhat different, although equally clever, approach of "thermophotovoltaics" may be able to yield efficiencies of 30 to 50 percent by shifting the solar spectrum to a range especially suitable for solar cells (Hammond 1977; Kelly 1978). Under this arrangement solar energy is absorbed and reradiated to a solar cell by a thermal body heated to about 3,300°F. Any light that is not initially used by the cell is reflected back to the radiating mass, thus keeping the energy within the system.

Economics

The price of silicon solar cells has dropped more than tenfold in the past five years, and in 1978 cells were available for $11 per peak watt[4] (Fan 1978). Average costs in 1979 were down to $7 per watt (McDowell 1979). A plethora of options exist for further reducing the costs of solar cells, many of which could bring prices down to $1 to $2 per watt within three to five years (Kelly 1978). One plant nearing completion in El Paso, Texas, expects to be able to produce arrays of thin film cadmium sulfide cells for a price of $2 to $5 per watt by 1980. Researchers at Westinghouse believe that production methods achieved in the laboratory by 1979 can soon lead to the availability of silicon ribbon cells at a price of 50 cents per watt; a

---

4. A cell rated at 1 peak watt will produce 1 watt of electrical power when exposed to conditions of maximum solar insolation, although the average output of such a cell would, of course, be much less than 1 watt, typically ranging from about one-fifth to one-quarter of a watt for installations in the United States. All subsequent costs cited in terms of dollars per watt in this section refer to peak watt costs.

research team at the University of Delaware predicts that technology currently under development can provide thin film cadmium sulfide cells for a price of 25 cents per watt by 1982 (SED 1978b). Another company is working on a process that may produce thin film cells for a price as low as 5 to 15 cents per watt (Kelly 1978).

Concentrating systems are already available at a cost below $7 per watt. For example, array costs for a concentrating system using silicon cells, under construction at the Mississippi County Community College in Blytheville, Arkansas, are only $2.75 per watt (SEIR 1978b). An estimate of only $1.80 per watt was made for solar-concentrating silicon arrays to be installed in a tentatively planned 10 MWe photovoltaic power plant in Minnesota (Frank 1978). Solectro-Thermo, Inc. (STI), a small company based in Dracut, Massachusetts, is currently marketing a thermal-electric system employing solar-concentrating collectors and silicon cells for residential and commercial buildings. Array prices for the STI system run about $6 per peak (electrical) watt. (Charlton 1979). The overall economics for this system are more favorable than this figure suggests, however, owing to the fact that useful thermal energy for space and water heating is also provided in adddition to electricity.

Cost goals for solar arrays (in 1975 dollars) set by the U.S. Department of Energy (DOE 1978) are $2 per watt in 1982, 50 cents per watt in 1986, and 10 to 30 cents per watt in 1990. Although the 1986 goal would represent a twentyfold drop from current costs, even larger cost reductions have been routinely achieved in the semiconductor industry. Semiconductor manufacturers and observers familiar with the industry are generally confident that the goal can be achieved without any breakthroughs, simply by the scaling up and automation of production (Hammond 1977). A study by the American Physical Society (APS 1979) recently concluded that flatplate silicon modules could be produced using presently foreseeable technology at a price of 50–75 cents per watt. The rate of cost reductions and efficiency improvements achieved for solar cells to date has in fact outpaced the goals established by the federal government (Hammond 1977).

Large-scale government purchases of solar cells would have a dramatic effect on costs by allowing for the expansion and automation of production capabilities. Several solar cell manufacturers recently concluded that government purchases of 30 MWe of solar cells at a total cost of $70 million, staggered over a five year interval, could bring prices down to $1 per watt (Frank 1978). Another study done for the Federal Energy Administration (1977) concluded that government purchases of 152 MWe of photovoltaic arrays to replace gasoline-powered electric generators at defense installations over a five year period could save almost $500 million in the long run and also lead to the availability of photovoltaic arrays at prices of 75 cents per watt by 1983.

As the costs of solar cells are reduced, the potential market for their utilization will expand accordingly. By the time photovoltaic arrays are available for a price of $1 to $2 per watt (expected in the early 1980s), a sizeable market will open up for irrigation and other remote water-pumping applications (estimated to be more than 6,000 MWe) and for the lighting of roads, parking lots, and other open spaces (Carpenter and Taylor 1978; Commoner 1978). At an array price of $1 per watt, photovoltaic-generated electricity would be economically competitive with diesel-fired electric generators (a total market of about 5,000 MWe) and at 75 cents per watt would also begin to compete with conventional electricity supplied to residential and commercial users in certain parts of the country (Carpenter and Taylor 1978; Commoner 1978). The achievement of the 50 cents per watt goal would lead to a tremendous growth in sales, because at this price, electricity from solar cells could in many cases compete with conventional utility power supplied to residential users throughout much of the country (OTA 1978; Carpenter and Taylor 1978). A resultant explosive expansion of production could in turn lead to even lower cell costs, making possible DOE's 1990 price goals of 10 to 30 cents per watt. At this level, solar electricity would be competitive for virtually all applications, including central station power generation (APS 1979; Kelly 1978).

Environmental Impacts

Photovoltaic generation of electricity is remarkably benign from an environmental point of view. The operation of photovoltaic devices produces no noise, pollution, or other wastes and requires no water. In many cases it will be possible to incorporate these devices into building structures, thereby minimizing the impacts associated with land use.

As with other solar devices, some pollution will result from the manufacture of solar cells if conventional energy sources are so employed. Present techniques for manufacturing high grade silicon are extremely wasteful in the use of both materials and energy, and consequently, the cells have to operate for about four years to deliver as much energy as was used in the production of silicon. However, production efficiencies can be considerably improved, so that it should be possible to manufacture cells that can produce all the energy used in their fabrication within four months (Kelly 1978).

This represents an extremely low energy cost in comparison to most other energy technologies.

Although silicon is nontoxic and extremely abundant, problems of materials availability are of some significance for solar cells composed with cadmium or gallium arsenide. Nonetheless, domestic supplies of cadmium will be capable of supporting annual production rates exceeding several thousand megawatts per year (Kelly 1978). Gallium is available in roughly comparable quantities and the arsenic supply poses no problem (Fan 1978). If gallium arsenide cells are used in solar-concentrating systems, requirements for gallium would be dramatically reduced. Although both cadmium and arsenic are potentially hazardous substances, proper cell encapsulation should minimize the toxicity danger (Fan 1978). Fires at residential installations, however, would pose a potential health threat. Nonetheless, any adverse impacts associated with photovoltaic conversion will be minimal and certainly dwarfed by the environmental advantages inherent to this approach.

Overall Assessment

Photovoltaic conversion offers virtually unparalleled benefits in comparison to other modes of power generation. Solar cells are inherently modular and thus can provide electric power on a variety of scales to fit given end use needs. These devices are quiet and nonpolluting and can be readily integrated into buildings, making them ideal for on-site applications. The ultimate potential of photovoltaics as a future power source is vast, because, apart from outstanding problems of storage and integration with present electrical systems, no major technical, environmental, or resource barriers to widespread utilization exist.

The major barrier to the extensive use of solar cells is their presently high cost. However, manufacturing cost reductions achieved to date have been encouraging, and experts familiar with the technology are generally confident that the cost of these devices can be brought down to competitive levels within a decade or so. Federal outlays needed to hasten large-scale commercialization will be smaller than that required for most new electrical generation technologies. The Photovoltaic Demonstration Act of 1978—passed by the 95th Congress and authorizing roughly $1 billion to be spent over a ten year period to reduce the costs of solar cells and lead to the early commercialization of photovoltaic applications—is surely a step in the right direction.

Numerous options exist for the design of solar cell arrays—including single crystal, polycrystalline, and amorphous silicon cells;

thin film devices; and solar-concentrating systems—and it is too early to predict which approach will ultimately win out. Imaginative concepts have been generated at an impressive rate, and great room exists for further technical innovation. It is possible that in the future solar cells may even be fabricated out of materials not presently under consideration. Therefore, a diversity of approaches to photovoltaic conversion should be encouraged in federal research and development programs until a clear preference for certain candidate systems can be firmly established. Although one cannot confidently predict which of the existing options will have the most favorable economics, some approaches do have clear advantages on other grounds. For example, the vast abundance and nontoxicity of silicon makes it a highly attractive material. The cost-cutting possibilities for amorphous silicon-hydrogen cells, which require lower quality silicon in greatly reduced quantities per cell area, are particularly intriguing. In addition, concentrating systems using highly efficient solar cells hold considerable promise. This approach is advantageous because it tends to minimize both land use and material requirements and also allows for the possibility of photovoltaic total energy systems, which should be attractive in applications for which a use for low temperature thermal energy exists.

Thus, despite uncertainties over the optimal approach to utilization, the overall prospects for photovoltaic power in the nation's energy future are good. However, the prospect can in part be blighted if the remaining difficulties facing widespread adoption of this technology are not forthrightly and vigorously addressed by the federal government.

WIND ENERGY

Introduction
The uneven heating of the earth's surface by the sun gives rise to the large-scale motions of the atmosphere referred to as the winds. The winds thus represent solar energy temporarily converted into the kinetic energy of air. This kinetic energy is continually replenished by solar radiation and continually dissipated by frictional processes that result in atmospheric heating.

Wind energy has been utilized by man for millenia. The Egyptians used wind to propel marine and river vessels 5,000 years ago. The Persians built windmills to grind grain more than 1,000 years ago (Merriam 1978). In Europe, the relative importance of wind reached a peak in the sixteenth century. With the coming of the Industrial

Revolution and the attendant availability of more convenient fossil fuels, the use of wind power declined dramatically.

Nevertheless, wind energy has made an important contribution in the United States. Over the last century, approximately six million small windmills have operated, of which roughly 150,000 are still in operation (Eldridge 1975). Until recently, the largest wind turbine constructed in the United States was a 1.25 MW machine installed on a hill called Grandpa's Knob in Vermont. Power generation began in 1941, two years after the initiation of construction, and continued through 1945, at which time a blade failure forced the plant to shut down. Wartime priorities prevented the replacement of the defective blade.

In recent years, there has been renewed interest in harnessing wind power, owing chiefly to higher fuel prices, an increased awareness of the environmental problems associated with conventional energy sources, and several attractive features of the wind resource. Wind represents a large and nondepletable energy resource that can be utilized with minimal impact on the environment, producing no air and thermal pollution and requiring no water in its utilization. The simplicity of wind technology will allow for rapid deployment in comparison to many other energy technologies. Finally, the economic prospects of wind systems are quite promising.

There are barriers as well to the widespread use of wind power, but none should prove insurmountable. Improvements are needed in the overall durability of wind generating systems and particularly in the ability of wind machines to withstand extremely rugged weather conditions. More information is needed on both the optimal size of wind systems for various applications and the optimal configuration for deploying large arrays of wind turbines. Finally, remaining obstacles to energy storage and to linking wind generators with conventional electricity systems need to be overcome.

Resource Potential

The maximum energy potential of the winds is uncertain, but appears to be large by all accounts. The global wind energy resource is necessarily small in comparison to the total solar energy input; it has been estimated that anywhere from about 1 to 4 percent of the total solar radiation incident on the earth's surface is converted to wind energy on a continual basis (Lorentz 1967; Lockheed 1976), with the current best estimate placed at about 2 percent (Gustavson 1979).

One estimate of the total power potential of the winds over the entire United States exceeds an average value of 10^{11} kW (NSF/NASA

1972) or, equivalently, about 3,000 quads per year. Only a minute fraction of this energy could practically be recovered, however.

If the rate of energy withdrawal from the winds exceeded more than a very small fraction of the amount naturally dissipated, there might be a significant modification of weather patterns. The fraction of wind energy that can be extracted without major impact is not accurately known at present. Gustavson (1979) has tentatively found 10 percent of the energy dissipated within 1 kilometer of the earth's surface (amounting to about one-third of the total wind energy dissipation) to be a prudent upper limit. In any event, the amount of energy that can be practically extracted will, more likely than not, be limited to a small fraction of the available potential by pragmatic technical and economic considerations. Specifically, the utilization of wind energy will be limited to physically accessible sites where near surface wind speeds are relatively high, averaging about fifteen miles per hour and where conflicting uses of land are absent.

In many regions, the density of wind power exceeds that of the incident solar flux. The reason why the wind energy flux can exceed the solar flux in certain locations is that the winds store energy. Thus, near the earth's surface, it may be possible to extract more energy per unit area than is locally generated by incoming solar radiation. Of course, on a sustained basis, one could never extract more wind energy than is continually replenished by the sun. At a highly favorable site, the annual flux of wind energy may attain an annual average value up to 46 W/ft^2 (500 W/m^2), which is twice as high as the average incoming solar flux at an especially good location (Merriam 1978). The annual average wind power density over the entire Great Plains region exceeds 18 W/ft^2 (200 W/m^2) (Metz 1977c), which is somewhat over the U.S. average insolation rate. Furthermore, wind energy can be converted to electricity with greater efficiency than can sunlight.

Estimates of the extractable resource potential vary considerably. At the lower end of the spectrum, one study concluded that 1 trillion kWh of electricity, or about 10 quads per year (primary energy equivalent), could be obtained from high wind areas of the United States on land that is not presently used or planned for use (Merriam 1978). Another study estimated that up to 20 quads of energy could be extracted annually, excluding off-shore regions (General Electric 1977). These estimates can be compared with the 2.2 trillion kWh of electricity (or roughly 23 quads of primary energy) consumed in the United States in 1978. Others have estimated, on the basis of putatively conservative assumptions regarding operating efficiencies and turbine spacing, that wind energy harnessed from the Great Plains

Map 4-2: Mean Annual Wind Power (watts/square meter) Over the United States. Wind power is estimated at a height of 50 meters above exposed areas. Over mountainous regions (shaded areas), the estimates are the lower limits expected for exposed summits and ridges. The map omits areas off the northeastern coast, where mean annual wind power may exceed 800 watts/square meter. (1 watt/square meter = 0.093 watts/square foot)

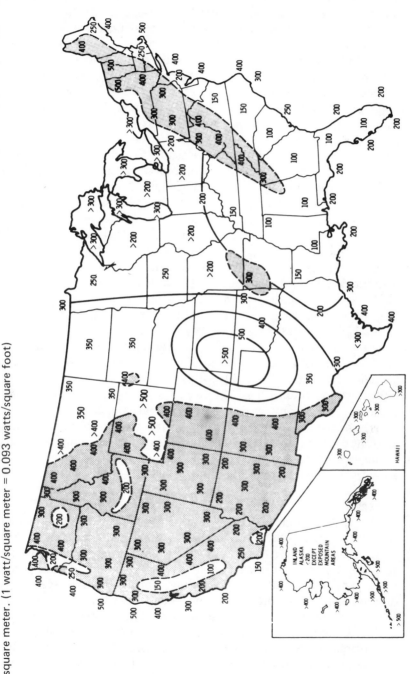

Source: D.L. Elliot, Pacific Northwest Laboratories.

alone could provide several times the present U.S. electrical energy use rate (Eldridge 1975; Blackwell and Feltz 1975). Regions that are also suitable for wind generation facilities include parts of Alaska, the Pacific Northwest coast, the New England coastal regions, the Great Lakes regions, and the Texas Gulf coast. Fortunately, many of the areas in the United States where the average wind speeds equal or exceed eighteen miles per hour at an altitude of 150 feet are near large population centers such as New York, Boston, Denver, San Francisco, and Los Angeles. Heronemus (1976), for example, has calculated that 9 quads per year could be generated from the Atlantic coastal area alone. Another study found that 6 to 9 quads per year of wind energy could be obtained in California (Craig et al. 1978). At the upper end of the spectrum, Gustavson (1979) has estimated the extractible upper limit to be 2 billion kW of continuous power, equal to 60 quads per year of electrical energy or about 180 quads per year on a primary energy equivalent basis. This estimate was arrived at assuming that energy extraction proceeded at a level equal to 10 percent of the rate at which wind energy is naturally dissipated within 1 kilometer of the U.S. land mass. Similarly, a study by Lockheed (1976) found the maximum wind energy potential from open range land in the conterminous United States to be about 150 quads of primary energy annually.

Thus, although considerable disagreement exists over the precise magnitude of the available wind resource, the overriding consensus is that it is large. Ultimately, it is probable that the overall contribution from wind will be limited to a greater extent by siting constraints, engineering and cost considerations, and an inability to integrate wind energy into our utility system than by the total amount of energy available. On the basis of the foregoing, we consider a range of 10 to 40 quads per year (primary energy equivalent) to be a relatively conservative estimate of the extractible wind resource.

Technology

The momentum of moving air can be transferred to that of a rotating wind turbine, and the intercepted mechanical energy can be used to pump water, compress air, generate heat, and, perhaps most significantly, drive electrical generators. A variety of configurations exist for wind machines, including horizontal axis systems with double, triple, and multiblade propellers and vertical axis systems. The power output varies as the square of the propeller diameter (for horizontal axis systems) and as the cube of the wind velocity, making it extremely advantageous to site wind machines in areas with high average wind speeds. The maximum percentage of wind power that is

Photograph 4-5: Horizontal Axis Wind Generator. Artists' conception of a wind generator designed to produce 2.5 megawatts of electrical power at a mean wind speed of 14 miles per hour. The generator, with a 300 foot long rotor supported on a 200 foot high tower, will be built by Boeing Engineering and Construction Company.

Source: Department of Energy.

theoretically extractible by an ideal horizontal axis machine is nearly 60 percent. The theoretical maximum for vertical axis machines varies with design. The highest efficiency yet reported for a real machine is about 50 percent, and wind generators can routinely convert 30 to 40 percent of the energy in winds into electricity (Merriam 1978; Metz 1977c). The fraction converted over the course of the year, however, is likely to be less than 30 percent, owing to the fact that the energy contained in winds below a design-rated wind speed cannot be fully captured. Some energy is also wasted when wind speeds exceed the design-rated limit.

Wind energy conversion technology is quite mature in comparison to other means of generating electricity from the various forms of solar energy. At the present time, the dominant technology is the two and three blade horizontal axis machine. A test unit rated at 100 kW has been operated sporadically by NASA since 1975. Other government-sponsored projects include a 200 kW machine, which has recently begun operation in Clayton, New Mexico, and a larger, 2 MW machine scheduled to become operational at Boone, North Carolina, in 1979. In addition to the federal program, a small commercial wind industry, consisting of about fifteen to twenty firms manufacturing small wind systems, presently exists in the United States (DOE 1978). Other companies are planning to make larger machines available.

Vertical axis wind generators have received comparatively little attention until recently, but offer some potentially significant advantages over the more conventional horizontal axis machines. The heavy generators can be located on the ground, so that lower cost towers can be used. The symmetry of vertical axis machines obviates the need for a mechanism to orientate the machine when the wind changes direction. In addition, they can operate in higher wind velocities than can conventional generators. The Darrieus rotor, with blades resembling an egg beater, is currently regarded as the most promising of the vertical axis wind turbines. Efficiencies up to 35 percent are considered attainable (Eldridge 1975). The largest system of this type, connected to a 200 kW generator, is now producing electricity in the Magdalen Islands of Quebec. If tests with this prototype Darrieus turbine are successful, a series of similar units will be installed on other islands within the region (Potworowski and Henry 1976).

An innovative concept for wind energy conversion has been proposed by Yen (1976). In this system, wind striking a cylindrical tower would be deflected through vanes, creating intense vortices, or miniature tornadoes, in the tower. The pressure difference between the center of the spiraling wind and the outside air would be used to drive a turbine. Preliminary results indicate that this system would be capable of generating more power than conventional wind generators of comparable size.

The optimum scale of wind generating systems is presently the subject of controversy. Although the U.S. government program has been geared almost exclusively toward the development of wind systems rated at about 2 MW or greater, some of the potential advantages of smaller scale systems have recently gained some recognition. These include potentially higher capacity factors, improved system reliability, and cost advantages stemming from a greater amenability to mass production.

The allowable density of wind machines on a given area is also uncertain at the present time. It is generally believed that so-called wake interference can be avoided by spacing wind turbines at least five to fifteen rotor diameters apart (Merriam 1978; Craig et al. 1978; Gustavson 1979). Less well understood, however, is the proper spacing limit for arrays of wind generators covering a large region (Merriam 1978). The basic problem is to avoid the overall depletion of the wind energy in the near-surface atmosphere, in order that undisturbed wind speeds would be available to machines located in central and downwind portions of the region.

Winds are inherently intermittent phenomena, although they are surprisingly reproducible on the average when viewed over periods of months or years. Nonetheless, the short-term variability of the winds makes energy storage necessary for most practical applications. One attractive option is to integrate wind-generated electricity into a utility grid having installed hydroelectric power capacity. The rate of hydroelectric generation is then regulated to take into account fluctuations in output from wind systems. Many regions of the United States, particularly the Pacific Northwest, have an abundance of hydroelectric installations in addition to being quite windy. Wind turbines could also be operated in conjunction with pumped hydroelectric storage facilities. Under this arrangement, wind power would be used to pump water "uphill" to an elevated turbine. Water would then be allowed to flow through turbines to produce electricity at a rate determined by the demand for power. One study conducted by scientists at the Interior Department's Bureau of Reclamation found that wind generators integrated with pumped storage facilities in western states could economically provide more than 100,000 MWe of generating capacity (equivalent to about one hundred large nuclear

Photograph 4-6:   Vertical Axis Wind Generator. This seven-story high vertical axis Darrieus wind turbine generator is being tested at Sandia Laboratories. The rotor, 56 feet in diameter, is capable of producing 30 kilowatts of electrical power in a 22 mile per hour wind, or 60 kilowatts in a 28 mile per hour wind.

Source: Sandia Laboratories.

plants), capable of producing electricity at a cost competitive with that from newly constructed conventional power plants (Commoner 1979).

Other energy storage systems currently under development, such as compressed air storage (Eldridge 1976a), advanced electrochemical batteries, and flywheels, have also been proposed for utilization with wind generation. They are discussed in Chapter 5. One study indicated that only ten hours storage capacity would make a Danish wind machine as reliable as a nuclear power plant (Sorenson 1976).

Owing to the variability of the winds, the capacity or load factors for wind generating systems are generally well below the 50 to 70 percent factors characteristic of conventional large power plants. Typical load factors for wind generators range from 15 to 25 percent, except in regions with exceptionally steady winds, where values of 35 to 40 percent or higher may be possible (Merriam 1977). Justus et al. (1976) found that at a height of 200 feet in the Great Plains

Photograph 4-7: Small-Scale Wind Generators: This wide-angle view shows several small wind generating systems being tested by Rockwell International at Rocky Flats, Golden, Colorado. Manufacturers of the systems are (from left to right): Elektro, North Wind, Zephyr, American Wind Turbine, and Grumman.

Source: Rockwell International Rocky Flats Plant.

and certain New England coastal regions, capacity factors of 60 percent can be achieved for 100 kW machines. In the same regions, a 1 MW turbine can operate at a capacity factor exceeding 20 percent. Capacity factors are considerably reduced at lower heights, dropping below 20 percent at a height of thirty feet for 100 kW machines at most sites.

The overall reliability of electricity generation can be improved and the required storage capacity substantially reduced by feeding electricity from numerous dispersed wind generators into regional electric power grids. One study found that dispersed installations of wind arrays amounting to several megawatts of overall capacity could provide power on a dependable basis 26 percent of the time (Metz and Hammond 1978). A study by Justus (1976) demonstrated that the availability of power from wind turbines can be increased significantly by deploying them in large arrays. It was shown that arrays of 500 kW generators could have capacity factors exceeding 40 percent in New England and the central United States. Energy storage capacity could increase reliability even further. For example, it was found that twenty-four to forty-eight hours of storage capacity could increase power dependability to 95 percent or greater in these regions. Storage requirements could also be reduced by interconnecting dispersed arrays of wind turbines with solar-electric facilities, since sunny, windless days and windy, cloudy days are common in many locations.

According to a net energy analysis performed by Lockheed (1976), a wind turbine could produce as much energy as was consumed in its production, including the mining of ores, installation, and operation within eight months or less. Heronemus (1976) has estimated the energy payback period for offshore wind power systems to be roughly twelve to sixteen months.

Materials availability is not likely to constrain wind power development. The large-scale deployment of wind systems might necessitate the expansion of steel production, but this appears to be the only materials-related problem and a highly tractable one at that (Lockheed 1976). In fact, wind power systems potentially have a net materials advantage over solar-electric options, because in addition to the higher attainable efficiencies, the typical propeller blade covers only about 10 percent of the area from which energy is intercepted (Metz 1977c).

Economics

The production of wind generating systems has not yet occurred on a scale large enough to permit precise estimates of potential capital costs and electricity generation costs. Current estimates of production costs for large numbers of small and large wind machines generally fall within a range of $400 to $1,000 per peak, or rated, kW (Eldridge 1976b; Metz 1977c; MITRE 1975). Electrical generation costs of 2 to 5 cents per kWh have been projected for large wind turbines on a mass production basis (Herman and Cannon 1976; DOE 1978). It has been estimated that electricity could be produced for 2 cents per kWh at especially favorable sites even if the capital costs of wind turbines were as high as $1,300 per kW (Metz 1977c). An example of such a site is the region around Medicine Bow, Wyoming, where the Department of the Interior's Bureau of Reclamation and the National Aeronautics and Space Administration have agreed to install a 1–5 MWe wind turbine, as announced in May of 1979 (SEIR 1979b). The wind generator, scheduled for operation in

1981, will be tied into the region's hydroelectric power system.

The uninstalled cost of the 200 kW Darrieus wind turbine in Quebec was quoted at $250,000 or, equivalently, $1,250 per kW. Large-scale production is expected to reduce the costs of these machines by about 35 percent, to roughly $800 per kW. WTG Energy Systems, Inc., of Angola, New York, has a 200 kWe wind machine available for an uninstalled cost of about $200,000 (i.e., $1,000 per kW). One of these machines installed in Gosnold, Massachusetts, supplies one-half of the town's electricity, at a cost competitive with electricity produced from the local diesel generating plant (Spaulding 1979). Wind Power Products, a small engineering firm from Seattle, Washington, is building a 3 MWe wind turbine for Southern California Edison for a contracted price (including installation) of about $350 per kW (Inglis 1978); operation is scheduled to begin in 1979. The price for subsequent units is expected to be only $200 per kW, leading to anticipated electrical-generating costs of 3 to 4 cents per kilowatt hour.

U.S. Wind Power, a private company from Burlington, Massachusetts, is planning to erect twenty 50 kWe windmills at Pacheco Pass, California, selling power to the state at a price of 3.5 cents per kWh, competitive with the cost of electricity from an oil-fired power plant (*New York Times* 1979). The first windmill is expected to be installed by 1980, and the twentieth should be on line by 1983. If successful, the company will install an additional 480 windmills at the site, bringing the total wind generating capacity to 25 MWe. The total project is estimated to cost nearly $19 million, or about $750 per installed kilowatt (SEIR 1979c). The California Department of Water Resources is also looking into the possibility of installing another 75 MWe of wind generating capacity at other locations around the state.

Numerous companies are also currently selling small wind generators ranging in size from 2 to 15 kW. Market prices for these machines average about $1,100 to $1,800 per kW (Metz 1977c).

The relative economics of competing wind systems of different sizes needs to be thoroughly investigated. Although analyses by the federal government and large industries generally indicate a cost advantage for large machines on the megawatt scale, this prediction has yet to be confirmed in actual practice (Metz 1977c). Small machines are required in larger numbers, provide a greater potential for manufacturing cost reductions, and can also be installed more rapidly than larger ones. Heronemus (1976) has estimated that small wind machines up to 100 kW in size can be mass produced at a rate of only $100 per kW, excluding the tower and support structure, which normally account for well under one-half the total cost. The feasibility of achieving these cost goals, however, remains to be demonstrated.

Although the ultimate costs of wind generation cannot be precisely predicted, it is evident that both small and large wind systems are promising from an economic standpoint. It has been estimated that a twofold reduction in costs could make wind power economically competitive with conventional electricity generation in many regions of the United States (Metz 1977c). Since materials directly account for only 10 to 20 percent of the present cost of wind systems, the potential for cost cutting is considerable (Metz and Hammond 1978). Small (25 kWe) wind machines are roughly comparable to small automobiles in terms of both weight and general complexity and are therefore highly amenable to mass production (Inglis 1978). Given that more than ten million cars are produced in Detroit each year, a wind power industry on a scale of only a few percent of that of the automobile industry could make a significant near-term contribution to the nation's electricity needs. In light of the limited scale of production of wind machines to date and the large possible cost reductions that could occur with the adoption of mass manufacturing techniques, the economic outlook for wind power is indeed highly attractive.

Environmental Impacts

The environmental impact of wind energy utilization should be minimal. Wind is in fact considered to be perhaps the most ecologically benign source of electric power (Harte and Jassby 1978). One particular advantage is that no water is required and no air or thermal pollution is produced. Weather modification effects are expected to be negligible if the amount of power extracted from the wind is only a small fraction of the natural rate that wind energy is dissipated into heat. There will be some environmental effects from the mining and processing of metals used in construction, but these are present in any power-generating system. Most of the other potential problems frequently cited are relatively insignificant. Wind turbines will generally be remotely sited, mitigating the possibility of noise pollution or visual offense. The land use impact should not be severe, because the siting of wind turbines will not preclude the possibility of other uses of land, such as agriculture. The hazards posed to migrating birds should be inconsequential compared to other sources of bird mortality (Merriam 1977). Television interference might prove to be troublesome in certain cases, although the precise loss that would be involved is open to debate. Regions affected by television interference are only expected to extend at most one mile from large wind machines; for small machines, interference is not expected to be a problem (Metz and Hammond 1978). Interference can be considerably reduced by using fiberglass, rather than metal, turbine blades.

Thus, in contrast to conventional methods of power generation, wind energy conversion is highly favorable from an environmental standpoint, posing no major societal risks.

Overall Assessment

Among the various solar-related electrical-generating options, wind energy technology is closest to readiness for widespread commercial introduction. The potential energy contribution from wind power is large and ultimately could provide a large fraction of total U.S. electricity requirements. Being no more complex than automobiles to produce, wind generators can be manufactured and installed quickly. Consequently, wind power can make a significant contribution in the near term, given a determined national effort. Unfortunately, the amount of attention and money being devoted to wind technology development by the federal government is astonishingly low given the technological maturity, economic attractiveness, and favorable environmental features of wind generation. This neglect could be corrected by an aggressive program geared toward the elimination of existing technical and economic barriers, in order that wind power can fulfill its great potential.

HYDROELECTRIC

Introduction

Hydroelectric power represents a form of stored solar energy that is, in principle, a renewable source of energy. In the course of the hydrologic cycle, water from lakes and oceans is evaporated by the sun. It is eventually returned to the land as rain. Impounded behind dams, it has stored gravitational energy. Allowed to flow through turbines, the water acquires kinetic energy. This energy is converted in the turbine to useful mechanical or electrical energy.

During the first part of this century, hydropower was used to produce from one-third to one-half of the electric power generated in the United States. This fraction has steadily declined as the use of coal, oil, gas, and, more recently, nuclear power has increased. Hydroelectric facilities currently provide about 13 percent of the U.S. electric-generating capacity and contribute a total of about 3 quads to the nation's energy supply. With most large sites already implemented, no large expansion of hydropower is possible. Consequently, hydropower cannot be regarded as a major source for new energy supplies.

Although the net production of electricity from hydropower cannot be significantly increased, the overall role of hydropower can be enlarged in other important ways. Existing facilities can be upgraded to increase the peaking power capability. Modifications of this nature would not add to the net supply of electricity, but would make it available at a faster rate when needed. The other major expanded new role for hydropower will be as a means of storing energy on a large-scale basis. Pumped hydroelectric storage facilities would utilize excess power to pump water "uphill." This gravitational potential energy could later be retrieved in the form of electricity when needed by allowing the water to fall back down through turbines. Pumped storage could become especially important in the future when used in conjunction with intermittent power sources such as the sun and wind.

Hydroelectric power has advantages and disadvantages in comparison to conventional electrical power sources. The principal advantage is that a renewable form of energy—falling water—is used and essentially no air or thermal pollution is produced. Hydropower plants may be started quickly and therefore are well suited for meeting peak power loads. The major drawbacks of large-scale hydroelectric projects are environmental in nature and relate to the deterioration of water quality and the considerable land use required for reservoirs. The environmental impacts associated with small-scale hydropower systems are much less severe.

Resource Potential

Installed hydroelectric capacity in the United States presently stands at slightly over 60,000 MWe. Estimates of additional hydroelectric power potential vary widely. Several government surveys have reported extraordinarily large values for potential future capacity (FEA 1974; Army Corps of Engineers 1975). These surveys generally represent merely inventories of possible sites. In general, the individual sites have not been systematically investigated to determine whether they are technically and economically feasible and environmentally sound and whether in fact there exist conflicting municipal, agricultural, industrial, or recreational demands for the available water. For example, in the Army Corps of Engineers' (1975) estimate that 4 additional quads per year of hydroelectricity could be made available, one-half of the rivers slated for the new generating facilities are included (or under consideration for inclusion) under the National Wild and Scenic Rivers System. A general conclusion that can be drawn from these surveys is that

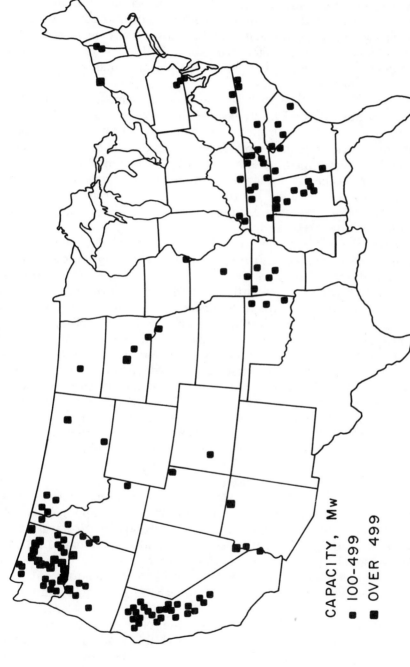

Map 4-3:   Developed Hydroelectric Resources in the United States.

CAPACITY, Mw

■ 100-499

■ OVER 499

Source: Federal Power Commission. "1970 National Power Survey." Washington, D.C.: Government Printing Office, 1971. p. I-7-12.

while total installed capacity, and hence peaking power capacity, can be increased substantially, the total amount of additional electricity generated will not be significant; only the rate at which electricity can be generated will be substantially increased. This also reflects in part the fact that many of the new facilities under consideration are to be pumped storage plants, which are net users, rather than producers of energy.

Large-scale facilities, such as the Hoover Dam, account for the major portion of total hydroelectric energy production today. One means of expanding hydropower is to construct more such facilities. But the majority of the most favorable hydropower sites have already been developed. Although it may be technically feasible to increase this capacity to some extent by the construction of new, large dams, it would be imprudent from an environmental standpoint. Therefore, we have excluded this as a possible option for expanding hydroelectric production.

Another possible means of increasing hydroelectric capacity is through the expansion and upgrading of existing hydropower facilities and the installation of electric generation equipment at existing large and small dams presently used for other purposes. A recent study by the Army Corps of Engineers (1977) estimated the maximum potential to be 54,600 MWe. Although this would represent a near doubling of total installed capacity, the net amount of additional electricity generated would be increased by only about 50 percent or a maximum net increase of 1.6 quads. It is doubtful whether this level of implementation would be technically feasible or desirable. The hydraulic head at one-third of the rivers in question may be too small for economical power generation (Huetter et al. 1978). The majority of streams where the dams are to be located also experience flow interruptions of one or more weeks duration. Furthermore, conflicting water uses may be incompatible with the practical requirements of power generation.

The installation of small-scale, "microhydro" systems on rivers and streams throughout the country offers another approach to increasing the net contribution from hydropower. However, this option has not yet been fully explored, and consequently, no definitive estimates of the power potential exist at present.

As an upper limit, we have concluded that an incremental contribution of 1 quad could be achieved by means of upgrading and retrofitting existing dam sites and by the construction of small-scale hydropower facilities where environmentally and economically justifiable. This estimate is consistent with that made by the Department of Energy (DOE 1978), but is considerably lower than estimates that have been made with little regard to environmental and economic constraints.

Technology

Hydroelectric power generation is a well-established technology. In conventional, large-scale hydroelectric plants, a large reservoir is created by a dam. Water collected during rainy periods and from spring runoff is stored and used for power generation throughout the year. In baseload plants, the power output remains relatively constant, whereas significant fluctuations of output occur with peaking units. The second major type of hydroelectric facility, referred to as pumped hydro storage, is discussed in the section on energy storage systems in Chapter 5.

Small-scale hydroelectric systems, using low head (less than about sixty feet in elevation) dams, are also technologically mature. The related industries are presently more advanced in Europe, where the use of these systems is much more prevalent than here.

Hydroelectric power can also be generated without dams. Small turbines may be installed directly in rivers, but a preferable approach would be to install the turbine within a special conduit or pipe through which water flow has been diverted. The so-called bulb turbine is likely to play a central role in nondam hydropower systems and may also be useful in the retrofitting of existing dams (Erskine 1978). This particular turbine is distinguished from other water turbines by virtue of its emplacement in a bulb-shaped housing that provides protection for the generator.

Strictly speaking, hydroelectric dam facilities cannot be considered long-term power sources, because the reservoirs will ultimately fill with silt, rendering them useless. This siltation process is normally completed within fifty to one hundred years, unless countervailing desiltation measures are undertaken. (Eckholm 1975; Odell 1975). Estimates of the useful life of dams and power generation equipment range from fifty years (Allen 1976) to one hundred years (Corso 1976).

Economics

Historically, hydroelectric power has been a relatively inexpensive source of electricity. The situation has changed, now that the most economically favorable sites have been developed, and the cost of power from future facilities will be much higher. While the costs of hydroelectric facilities completed in 1971–1972 ranged from $93 to $726/kWe (FPC 1975), some future projects are estimated to cost up to $2000/kWe (*Electrical World* 1976). Due to the long lead times

Photograph 4-8: High-Head Hydroelectric Generating Station. The Curecanti Unit of the Morrow Point Dam on the Gunnison River near Cimarron, Colorado generates 120 megawatts of electric power.

Credit: V. Jetlay.
Source: Bureau of Reclamation Denver Office.

Photograph 4-9:   Low-Head Hydroelectric Generating Station. The Wareham, Massachusetts generating station is being restored by the town of Wareham with a grant from the Commonwealth of Massachusetts. When operating, the station will generate 250 kilowatts of electric power.

Source: Phyllis Gardiner, Massachusetts Office of Energy Resources.

required to build large-scale hydroelectric plants, the cost of capital is a major contributor to the high construction costs. Reservoir preparation—particularly if large areas of inhabited land must be cleared—along with electric transmission and environmental protection, can add significant costs as well.

The economics become substantially altered, however, when a dam already exists. Upgrading and retrofitting existing dam sites is often economically attractive, requiring a relatively low capital outlay per kilowatt of capacity in comparison to other means of electricity generation (Erskine 1978).

Environmental Impacts
Although hydroelectric power generation produces no air pollutants or other waste by-products, major impacts are associated with large-scale dam projects. These problems are mitigated, to some extent, in small-scale, low head hydropower systems and are largely avoided in nondam systems.

Water quality can be severely impaired by the damming of free-flowing streams. The evaporation of large quantities of water occurs in reservoirs, increasing the concentrations of salt and other minerals in the remaining water. This is a serious problem for the Colorado River and other mineral-rich waters. Water quality is further degraded by nitrogen supersaturation, which occurs when water plunges below the surface at the bottom of spillways below dams. While a free-flowing stream will aerate and cleanse itself, a stagnant body of water will not. Temperatures of the impounded water also generally increase. The increased temperature, combined with the reduced aeration of reservoir water, can lead to changes in aquatic plant and animal life, frequently culminating in algae blooms. The dying off of algae can accelerate the depletion of oxygen dissolved in the water in a process called eutrophication.

Altering the natural flow of a river can have a profound effect on both aquatic and nonaquatic organisms. Dams pose a physical barrier to fish migration, and the formation of reservoirs may eliminate native fish-spawning grounds. The population of fish species often changes, and, in a number of reservoirs, so-called "trash" fish come to dominate. The creation of reservoirs can also cause significant ecological impacts and can adversely affect terrestrial wildlife by the disruption or elimination of natural habitats.

The construction of dams can have an adverse effect on man as well, in some cases causing the massive relocation of people. Dam failures can also take a heavy toll in human life and property, as witnessed in the 1973 Rapid City, South Dakota, disaster. In this accident, the Canyon Lake Dam—used for flood control rather than for power generation—gave way under heavy rains, killing 236 people (Odell 1975). Large dams may also cause earthquakes, with potentially disastrous results. Finally, the disappearance of natural, unspoiled rivers and the attendant loss of esthetic benefits and recreational opportunities (such as whitewater boating) is a significant cost born by society that should be carefully weighed.

Overall Assessment

The overall energy contribution from hydroelectric sources cannot be substantially expanded in the future. The most suitable sites have already been developed, and a combination of economic, environmental, and social considerations militate against the development of new, large-scale facilities. The major expansions that can be made will increase peaking power and energy storage capabilities substantially, without significantly adding to the overall amount of electricity generated. Small but important incremental contributions can be made in the future by upgrading and expanding existing hydro-electric facilities and through the retrofitting of existing large and small dams for the installation of electrical generation equipment. This approach generally looks attractive from both economic and environmental standpoints. Small, local contributions may also be made by lowhead, "microhydro" systems, although the prospects for this approach are uncertain and appear limited at best.

BIOMASS

Introduction
Photosynthesis is the biological process by which green plants convert sunlight into organic matter. The biomass resource consists of the chemical energy stored within organic matter originally derived from the photosynthetic process. The average energy content of raw, dry biomass is about 15 million Btu per ton, comparable to that of western coal (Hammond 1977). Roughly 0.1 to 0.2 percent of the solar energy falling on the earth is captured by plants. Although this percentage is small, the world biomass energy resource is quite large; on an annual basis, biomass production is about seventeen times greater than worldwide nonfood energy use (Poole and Williams 1976).

In many developing countries, wood, charcoal, and cow dung are the primary fuels used in rural areas. As recently as one century ago, wood provided the majority of the energy used in the United States. After hydroelectric energy, biomass is the largest source of solar energy currently utilized in this country, accounting for an estimated 1.3 to 1.8 quads per year or roughly 1.7 to 2.3 percent of the total (CEQ 1978; DOE 1978). This use mainly consists of the burning of residues in the wood products and paper industries and wood burned for heating in residential stoves. More recently, small amounts of ethyl alcohol (ethanol) derived from the fermentation of grain, corn, or other agricultural products have been used to extend modestly gasoline supplies in certain parts of the country. The mixture, called "gasohol," consisting of 90 percent gasoline and 10 percent alcohol, is now being sold in over 800 gas stations in at least 28 states (DOE 1979a). In the future, biomass is likely to play a vital role by supplying limited amounts of liquid transport fuels, methane, and carbon feedstocks to chemical industries. The magnitude of this contribution, however, is difficult to predict and may prove to be relatively small in comparison to total U.S. energy needs owing to fundamental limitations imposed on the biomass resource.

Biomass is an attractive energy source because the plants themselves naturally collect and store solar energy, functions that normally dominate the costs of most other solar conversion technologies. Unlike fossil fuels, the burning of plant matter or derivative fuels does not contribute to a net increase in atmospheric carbon dioxide, because on the average carbon dioxide would be consumed in photosynthesis at about the same rate as it would be released from combustion (Poole and Williams 1976). A major limitation of photosynthesis, however, is its relatively low solar energy conversion efficiency in comparison to other solar technologies. Typical efficiencies for U.S. agriculture range from 0.25 to 0.75 percent (von Hippel and Williams 1975). Maximum practical values are estimated to fall within the range 1 to 4 percent (AIAA 1975; Poole and Williams 1976). As will be discussed later, a major consequence of the low efficiency is that large areas of land would be required to produce large amounts of energy from photosynthetic conversion.

Biomass resources can be divided into two general categories—plantations that produce energy crops and organic residues and wastes. The term plantation refers to an area that is used to grow biomass strictly for energy purposes, whereas organic residues and wastes include such sources as agricultural crop residues, logging residues, urban refuse, and manure wastes.

Resource Potential

Plantations. The average photosynthetic production of all biomass in the United States, including agricultural crops and forests, is approximately three and one-half tons of dry matter per acre per year, corresponding to a solar conversion efficiency of about 0.25 percent (Poole and Williams 1976). Agricultural production typically yields about three and one-half to ten tons per acre per year. Some plants are capable of converting sunlight to chemical energy at efficiencies much greater than average. For example, sugar canes, red alder, and eucalyptus have high growth rates ranging from about ten to twenty-five tons per acre per year (AIAA 1975; Poole and Williams 1976). Favorable climates and special growing conditions, however, are required to obtain these yields.

The availability of land and water will impose serious constraints on biomass energy production. Most of the prime, arable land in this country is already used for agricultural production (Pimental et al. 1976). Much of the land that could be cultivated in the future is of marginal fertility, requires extensive irrigation, and would in any event be better utilized for the production of food, which is a much more valuable commodity per pound than fuel. Furthermore, over thirty-five million acres of productive cropland have already been lost to highways and urban development (Pimental et al. 1976).

A simple example will illustrate the large amounts of land and, more significantly, water that would be required for biomass plantations. Presently thirty-five million acres of agricultural land—equal to about 1.5 percent of the U.S. land area—are being irrigated in the United States (Groth 1975). Assuming that an additional thirty-five million acres were irrigated for the cultivation of energy crops with average yields of fifteen tons per acre per year and converted to methanol at 40 percent efficiency, a total of about 3 quads per year of usable energy could be produced. Considering that about 500 tons of water are typically needed to produce a ton of plant matter (Poole and Williams 1976), a total of 260 billion tons of water would be consumed for irrigation in a single year. This amount of water is equivalent to nearly 14 percent of the total annual stream runoff in the United States (Forest Service 1975). If only one-fourth of this land area were instead covered with 40 percent efficient solar collectors, about 80 quads per year, slightly more than the present national energy use rate, could be produced. If all of the cropland currently under cultivation in the country, about 330 million acres or 14 percent of the total U.S. land area, were used for the production of fuel alcohols, less than 30 quads could be provided. If all of this land were to be irrigated, the amount of water required would exceed the total annual stream runoff by about 33 percent. Clearly, biomass plantations could never supply a major portion of our total energy requirements.

Poole and Williams (1976) have suggested that large areas of land— an estimated 100 to 150 million acres, chronically subject to wetness problems and therefore unfit for agriculture, could be used for the production of biofuels. However, the practicality and environmental consequences of effecting this are unknown. Those areas designated as wetlands, considered to be prime wilderness areas and rich habitats for wildlife, should certainly be preserved. The value of marshlands has been estimated by one source to be perhaps greater than that of even the richest farmlands, owing to their functions in oxygen production and wildlife support and their potential use for water treatment (*Audubon* 1976). Whether the development of other waterlogged areas can occur without severe impact should be investigated.

In addition to cultivated cropland, there are 500 million acres of commercial forestland in the United States. The energy content of the total annual growth of salable timber is about 4 quads (Forest Service 1974, 1975). The total annual growth of all biomass in commercial forests is about 7 to 8 quads (C&EN 1979). If converted to fuel alcohols, the total usable energy obtainable from the annual growth of timber and total forest biomass would be only about 2 and 4 quads per year, respectively. For obvious reasons, collection of all the stump, branch, and shrub growth reflected in the latter figure would be highly impractical.

On a land use basis, timber growth represents an extremely inefficient energy source. Given present average timber growth rates, more than 200 million acres of land (nearly 9 percent of total U.S. land area) would be required to producd 1 quad of fuel alcohol. Even if it is assumed that the total annual growth of biomass is collected and utilized, a considerable amount of land—in excess of 100 million acres—would still be required.

The total energy content of the 4.5 billion cubic feet of annual mortality (Forest Service 1974, 1975) is only about 1 quad, or about 0.5 quads after conversion to fuel form. This resource is essentially unavailable, because trees in this category are so widely distributed that collection is virtually impossible. Nonetheless, the dead trees play a vital role in the forest ecosystem, and the fact that they are left in the forest does not mean that they are wasted.

Thus, even in theory, the energy potential from commercial forests is quite limited. In practice, there are already indications that many private and public timberlands are being overcut (*Sierra Club* 1975; Green and Setzer 1974; Forest Service 1974), even without the additional strain that would be imposed if forests were to be utilized for energy purposes. Furthermore, it does not appear that substantial increases in annual yields can be practically achieved. The major options proposed, such as the expansion of clear cutting or the use of fertilizers, would be unacceptable on environmental grounds due to excessive soil erosion and water pollution. In addition, future requirements for lumber and paper products are almost certain to increase in the future. Therefore, even if the nation's future timber requirements can be met by the expansion of logging activities, it is clear that forests cannot be regarded as a major source for biomass energy.

The cultivation of marine or fresh water biomass has also been proposed. The dominant concept under consideration is a large-scale, ocean-based kelp farm. Annual kelp growth rates of five to ten tons per acre have been reported (AIAA 1975). Assuming the latter figure, seventeen million acres of ocean surface would be required to produce 1 quad of fuel energy. This is equivalent to about twenty-six thousand square miles, or an area slightly larger than the state of Vermont. In addition to the extensive areas required, other potential problems exist. Kelp growing would consume large quantities of phosphorus, an essential nutrient. This must be provided by pumping phosphorus-rich water from the ocean depths to the kelp. Poole and Williams (1976) have computed that the total annual world production of phosphorus would support the production of less than 28 quads annually of methane gas from floating kelp farms. They also note that the upwelling of large amounts of nutrient-rich deep ocean water to the surface could add large amounts of carbon dioxide to the atmosphere; roughly three times more carbon dioxide would be released from the production of 1 Btu of methane from kelp than from the burning of 1 Btu of a fossil fuel. The pumping of large amounts of cold water to the ocean surface could also affect weather patterns. The possible interference of large seaweed plantations with coastal shipping also needs to be examined.

The cultivation of aquatic algae has also been suggested. Poole (1974) has estimated that growth rates would range from one-half to eighteen tons per acre per year, resulting in the large area requirements typical of other energy crops. Another problem is that bacterial digesters have to be maintained at relatively high temperatures in order to convert algae into methane, requiring an amount of energy equal to about one-half that of the generated methane (Poole 1974).

Therefore, neither seaweed nor microalgal biomass plantations appear to be practical at the present time. Consequently, we have excluded them from consideration as possible biomass energy sources.

Residues and Wastes. Organic residues and wastes are normally classified into four major categories—agricultural crop residues, forestry residues, urban refuse, and manure wastes.

Agricultural residues total about 340 million tons per year (Poole 1975; Sarkanen 1976). Much of this could be collected during harvesting and converted to fuels. After conversion, the nutrients contained within the leftover residuum should be returned to the soil as a fertilizer. If commercial fertilizers were used instead, significant amounts of energy would be consumed in their production, thereby undermining the gains achieved through bioconversion.

Considerable disagreement exists over the magnitude of the forestry waste resource. A middle-range estimate of 300 million tons of dry wood wastes has been made by Poole and Williams (1976). Included within this category are roughly one hundred million tons of logging residues (Forest Service 1973), another one hundred million tons of residues from wood processing mills and wood products plants (Forest Service 1974), and about fifty million tons of pulp wastes from the paper industry (Sarkanen 1976).

The amount of refuse produced annually in urban areas is about 135 million tons (Kasper 1974; Poole 1975). Additional refuse is produced in rural areas but would be difficult to collect in significant quantities.

The total amount of dry manure from all cattle is approximately 200 million tons per year (Feldman et al. 1973; Sarkanen 1976). However, only a fraction of this, about forty-five million tons, is produced annually in confined feedlots where it can be easily collected (Poole and Williams 1976).

Overall estimates of the energy potential for organic residues and wastes are presented in Table 4-2. We have assumed the energy content of all wastes to be 15 million Btu per ton, except for urban refuse, given a lower value of 9 million Btu per ton. To arrive at lower bound estimates, we have estimated that 50 percent of the potential resource is collected and subsequently converted into fuels such as methanol or methane with an average efficiency of 40 percent. For a more optimistic estimate, we have assumed the collection of 80 percent of the potential wastes and an average fuel conversion efficiency of 50 percent. As a maximum upper limit, we have assumed the collection of 90 percent of the organic wastes and their conversion to fuel with an efficiency of essentially 100 percent by means of a process proposed by Antal (1976) and discussed later in the chapter.

An intriguing proposal for the production of alcohol fuels and methane gas that does not strictly fall into the biomass plantation or residues and wastes categories has been made by Commoner (1979). This proposal involves the integration of biomass fuel production with livestock production. Under this arrangement, crops such as corn, sugar beets, and hay, grown to support livestock, are first fermented to produce ethyl alcohol (ethanol). The residue from the fermentation process is fed to livestock. Livestock manure is then collected and converted to methane. Finally, the residue from

Table 4-2. Energy Potential of Organic Residues and Wastes.

Category	Total Resources (in million tons)	Energy Potential (in quads)	Net Useful Energy (in quads)
Crop Residues	340	5.1	1.0 -2.0- 4.6
Forestry Residues	300	4.5	0.9 -1.8- 4.0
Urban Refuse	135	1.2	0.25-0.5- 1.1
Manure Wastes	45	0.7	0.15-0.3- 0.6
Total	—	11.5	2.3 -4.6-10.3

methane production is returned to the soil as a fertilizer, closing the cycle. It has been claimed that 8 to 10 quads of biofuels could eventually be produced in this manner.

While the general concept is certainly deserving of further study, the magnitude of the contribution that could ultimately be made through this approach is far from certain. The extent to which acreage devoted to growing livestock crops would have to be expanded to support the same level of livestock while simultaneously producing fuels remains an open question. Another fundamental problem stems from the fact that the production of ethanol is a highly energy-intensive task that may even require more energy than is given off when the alcohol fuel is burned. This problem is discussed in the next section.

Technology

Organic matter derived from energy crops, wood, and organic residues and wastes can be utilized in many ways. The simplest method of converting organic matter to usable energy is to burn it directly. However, the biomass resource could be used to better advantage by converting it to liquid and gaseous fuels and chemical feedstocks, applications for which it is uniquely suited. Plant biomass can be converted to a number of different fuels.

Methane gas can be produced by the bacterial decomposition of organic matter in an oxygen-deficient atmosphere in a process called anaerobic fermentation. Biogas digesters are in widespread use in China, with an estimated seven million units currently in operation (Smil 1977). A 50 percent conversion efficiency is considered possible, but improvements beyond that are uncertain (Poole and Williams 1976; C&EN 1979). A residual sludge that is rich in nutrients is formed during the decomposition and can be returned to the soil as a fertilizer. The precise fertilizer value of the organic residuum from biological decomposition is somewhat open to question, however, given its low nitrogen content and also the fact that it contains copious quantities of water, which might pose problems for handling and distribution (Antal 1979).

Ethyl alcohol or ethanol can be produced by the fermentation of corn, sugar beets, wheat, and other food grains. Many analysts have concluded that it presently takes at least as much, if not more, energy to produce a gallon of ethanol by means of fermentation of grains and the subsequent distillation of alcohol than the energy that can later be made available by burning it (McQuiston 1979; Chambers et al. 1978). This issue is highly controversial, however. A recent study

Photograph 4-10: Biomass to Gas Conversion. This prototype unit, located at the University of California-Davis, uses farm and forest residues to produce methane gas. The gas from this system fires a boiler that heats and cools one of the campus buildings at a cost comparable to that of other energy sources.

Credit: J. Schneider.
Source: Department of Energy.

by the Office of Technology Assessment (OTA 1979) indicates that ethanol can be produced with a favorable energy balance.

Methyl alcohol or methanol (commonly known as wood alcohol), on the other hand, can be produced from generally less expensive sources including wood, garbage, manure, and agricultural wastes such as corn stalks. In order to make methanol, organic matter is heated to a high temperature and converted to a gas composed of carbon monoxide and hydrogen. Methanol can be formed by passing the gas over a suitable catalyst. Methanol may be produced by several other methods as well (Reed and Lerner 1973; Reed 1976a). At the present time, wood and other organic matter can be converted to methanol with a net efficiency of 35 percent. An efficiency of 45 percent is considered to be a probable upper limit (Poole 1974).

Pyrolysis involves the heating of organic matter in an oxygen-depleted atmosphere to produce a range of fuel forms including liquid methanol, a solid char, and other liquid and gaseous fuels. Antal (1976) has suggested an interesting variant on the pyrolysis concept. He has proposed the use of high temperature solar heat for the conversion of organic matter to hydrogen in a steam boiler. An overall efficiency of 70 percent, including the solar energy input, has been predicted for this process. This general technique could also be used to produce hydrocarbon fuels such as methanol and gasoline. The principal virtue of this approach is that through the efficient use of solar energy, the inherently limited energy potential of biomass residues and wastes can be significantly expanded. The potential benefits are sufficient to justify a thorough investigation into the feasibility of this concept.

Organic matter can also be converted to organic compounds such as benzene, ethylene, and phenols, which can be used as feedstocks for plastic and chemical industries (National Research Council 1976). Feedstocks to chemical industries may also be derived on a sustained basis by tapping trees that produce oleoresins (naval stores and gum resins)—a practice that reached a peak in the United States near the beginning of this century (C&EN 1979). After trees have passed their prime for resin production, they may be harvested as wood biomass.

Organic matter constitutes approximately 60 percent of urban refuse. Normally, the nonorganic matter is separated prior to processing, and material such as aluminum, copper, glass, and steel can be recycled. The most commonly employed methods of utilizing urban refuse today are direct combustion and pyrolysis. As of 1979, there were twenty garbage-to-energy plants operating in the United States (Hirshorn 1979). New York City presently has major plans to build plants to convert its garbage—generated at the rate of about

Photograph 4-11:   Garbage to Energy Conversion. This, plant, located in Saugus, Massachusetts, burns trash from surrounding towns to produce steam which in turn is used to generate electricity for a nearby manufacturing plant.

Source: Department of Energy.

20,000 tons per day—into electricity and other forms of energy. According to an analysis by the Urban Development Corporation, the garbage could supply between 5 and 6 percent of the city's energy needs (Dionne 1979).

Economics

The economics of most biomass energy systems are uncertain. Hammond (1977) has suggested production costs ranging from $4 to $5 per million Btu as a probable range for most biomass fuels. As a point of comparison, average 1978 costs for regular gasoline, residential heating oil, and natural gas were respectively about $3.30, $2.30, and $1.80 per million Btu (DOE 1979b). Total production costs for methane may be $2 to $4 per million Btu (Poole and Williams 1976). Reed (1976b) has estimated the total cost of converting wood to methanol to be slightly under $6 per million Btu.

Ethanol is approximately three times more expensive to produce than methanol (Schuon 1979). Antal (1976) predicted that hydrogen could be produced at a cost of $3 to $5 per million Btu. Wood waste converted to pellet form may cost only $1–2 per million Btu (C&EN 1979).

The overall economics of organic residue and waste utilization is enhanced by the added value of by-products such as recovered metals and residual fertilizers and by the reduced costs of refuse disposal. Several plants in commercial operation today have demonstrated the economic feasibility of urban refuse utilization. In the long run, the costs of utilizing organic residues and wastes may prove to be largely immaterial owing to the unique potential of these materials as renewable sources of methane, liquid transport fuels, and chemical feedstocks and to the potential benefits stemming from a reduction of solid waste volumes.

Environmental Impacts

The environmental impacts of large-scale biomass energy plantations would be significant and similar to the impacts currently caused by agriculture. These effects include alteration of the earth's heat balance, radical changes in natural vegetation, heavy water use, water contamination from soil and fertilizer runoff, and the deterioration of soil quality. As has been demonstrated, limitations on the availability of land and water will ultimately preclude the implementation of this approach on a very large scale.

The utilization of organic residues and wastes will, on balance, yield net environmental benefits by reducing water pollution and land use for waste disposal. However, certain conditions must be satisfied if this is to be done in an environmentally acceptable manner. For example, the collection of agricultural crop residues represents the removal of nutrients from the soil. These essential nutrients must ultimately be returned if soil quality is to be maintained. It is not yet known what fraction of crop residues would best remain uncollected to avoid the depletion of nutrient and mineral content of the soil. The extent to which by-products from biofuel conversion can substitute for commercial nitrogen fertilizers also remains to be determined. For similar reasons, a sufficient quantity of organic matter must be left behind in the forest to avoid soil depletion. Care must be exercised in logging practices to prevent the overcutting of forests and to mitigate the problems of soil erosion and water contamination that have occurred in the past.

Overall Assessment

In conclusion, biomass energy should play a vital role in supplying methane, liquid fuels for transportation, and carbon feedstocks to chemical industries, although the precise magnitude of that role is subject to major uncertainty. Energy production from biomass plantations will be severely constrained by land and water availability. These constraints can be alleviated to some extent by improvements in both photosynthetic conversion efficiencies—leading to enhanced crop yields—and biofuel conversion efficiencies—resulting in a greater utilization of the biomass resource. The practical feasibility and environmental acceptability of growing energy crops on lands subject to chronic wetness problems warrants further study as a means of bypassing land and water limitations. Marine biomass systems should be ruled out until numerous environmental and engineering questions can be resolved.

Although it would be imprudent to devote large amounts of water and fertile land exclusively to the cultivation of energy crops, the integration of biomass energy production with agricultural, wood, and fiber products industries should be encouraged. Central to this concept would be the utilization of organic residues and wastes that constitute a valuable, although inherently limited, resource. The proper utilization of this resource could alleviate the environmental problems associated with solid waste disposal and simultaneously make an important contribution to the nation's energy needs. A major effort should be devoted to developing high efficiency processes to convert collected organic matter into useful fuels and chemical feedstocks. If environmental studies indicate that a large fraction of the potentially available wood and crop residues can be utilized without seriously degrading soil quality, biomass residues and wastes may play a sizeable role in providing liquid fuels for a future transportation sector.

SOLAR SATELLITE POWER STATIONS (SSPS)

Introduction

The idea of collecting solar energy in orbiting satellites thousands of miles from earth and transmitting it by microwaves back to a land-based receiving antenna was first proposed more than a decade ago. The concept is now beginning to attract serious consideration, and in 1979, Congress was entertaining proposals to increase federal investments in solar power satellite development from $4 to $30 million per year. Conceptually, solar satellite power stations represent one of the more visionary approaches to the generation of electricity from solar energy. However, a number of practical problems associated with this option—particularly the high costs of development and unanswered questions pertaining to potential environmental and health hazards—appear sufficient to rule out its introduction in the near term and also cast doubt on its future implementation.

Technology

The basic concept involves the collection of solar energy on large satellites in geostationary orbit and the subsequent generation of electricity by means of photovoltaic or thermal-electric conversion. The electricity would be converted to microwave form and transmitted to a receiving antenna ("rectenna") on earth, where the microwave energy would be reconverted to electricity and fed into the electric power grid.

Photograph 4-12: Solar Satellite Power Station. Artist's conception of a Solar Satellite Power Station. After being constructed in space and placed in a geosynchronous orbit, the satellite might beam 5,000 to 10,000 megawatts of power to earth-based receiving stations.

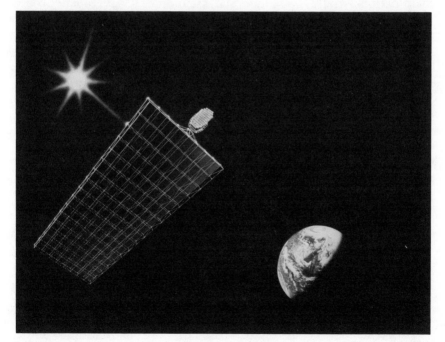

Source: National Aeronautics and Space Administration Johnson Space Center.

Present designs call for systems capable of delivering 5 gigawatts (GW) of electric power—equivalent to more than 60 percent of the peak power demand of New York City. Very large satellites, weighing approximately forty to eighty million pounds and having about twenty to forty square miles of collector panels (Glaser 1977)—an area roughly the size of Manhattan—will be required. It is anticipated that the satellite would be assembled from components in a relatively low orbit, about 200 miles above the earth's surface, and then "towed" to an altitude of about 22,000 miles. At this altitude, the satellite would remain in a fixed (geostationary) position relative to the earth, orbiting at the same rate as the earth's rotation.

The principal advantage of this approach is the increased availability of sunlight in space over terrestrial conditions. Sunlight would be available 99 percent of the time, offering the potential for virtually continuous generation of electricity. However, this advantage is more than offset by the numerous problems involved in the actual operation of an SSPS system.

Due to the necessarily large scale of the equipment involved and the great remoteness of the collectors, SSPS is an inherently inflexible and vulnerable form of power generation. The consequences of malfunctions would be severe, and maintenance and repairs would generally be complex and expensive. Many advances over current technology will be required to transform the SSPS concept from vision to reality. Significant improvements will be required in space transportation: a higher capacity version of the space shuttle will be needed to ferry construction workers, and heavy lift launch vehicles will have to be developed for the transport of construction materials (Witkin 1979). Mastery of the complex in-space construction techniques will require years of testing and development. Although the technical demands for reliable power generation from SSPS are exacting, the most significant associated problems are of an economic and environmental, rather than technical, nature.

Economics

The generation of electricity in space will undoubtedly be an expensive proposition. Capital costs for SSPS will be high and are currently projected at about $2,400 per kWe (Witkin 1979), with other estimates ranging from $1,500 per kWe (Glaser 1977) to $2,700 per kWe (DOE 1978). The estimated cost of electricity produced by the system ranges from 2 to 15 cents per kilowatt hour (Witkin 1979). These cost figures are not based on the construction of only a few satellites but instead depend on a high level of deployment, requiring up to one hundred satellites and capable of providing a large fraction, if not all, of the present U.S. electrical needs. However, even these cost estimates might be optimistically low.

Not included in the capital cost projections are the huge costs needed for space station and space vehicle technology development. Glaser (1977) estimated these costs to amount to about $44 billion, whereas government estimates of the research and development funding level needed to accomplish full-scale demonstration range from $50 to $85 billion (DOE 1978).

Furthermore, there is no convincing evidence to date demonstrating a decisive cost advantage for space-generated electricity over terrestrial solar-electric systems. Development costs for SSPS will certainly exceed those for land-based solar by a large margin, because the former requires major advances in space transportation and space technology in addition to improvements in the more "mundane" photovoltaic or solar-thermal conversion technologies.

Environmental Impacts

The environmental impacts associated with a large-scale deployment of SSPS would be potentially severe and warrant thorough examination. Putting the power stations into orbit would involve major problems. It has been estimated that 360 Space Shuttle flights would be required for the construction of a 5,000 MWe station (AIAA 1975). A more recent study estimates that sixty to one-hundred heavy lift launch vehicles (larger than the shuttle) would be required per satellite (Glaser 1977). Regardless of the vehicle type actually employed, large amounts of rapidly diminishing liquid fuels would be required. The net effect of the pollutants that would be introduced into the atmosphere from the launch vehicles is presently unknown, but several possible complications exist. Large quantities of water vapor, the principal exhaust of rocket engines, would be introduced into the upper atmosphere, a region where the water vapor content is normally low. The climatic impacts are difficult to predict, but the global energy balance might be affected by atmospheric heating or increased cloud cover. Nitrous oxides and chlorine gas from rocket exhaust would have a deleterious effect on the stratospheric ozone layer (Schneider and Mesirow 1976; von Hippel and Williams 1975), although the magnitude of this effect is difficult to assess at present.

Orbiting satellites would collect and transmit energy that would not otherwise be incident on the earth. In the absence of engineering measures designed in some way to reflect incident solar radiation from the rectenna or adjacent area, extensive energy production from SSPS would contribute in a minor way to a net warming of the earth.

The potential health hazards associated with microwaves also deserve careful consideration. The energy density at the center of the microwave beam would be about ten times higher than the U.S. standard of 100 W/m^2 for allowable long-term exposure (AIAA 1975). A growing body of evidence indicates that biological damage may result from exposure far below current U.S. standards (Brodeur 1976). The problem is not the possibility of the microwave beam wandering astray and accidently "roasting" a hapless residential area, which would be prevented by safeguard mechanisms, but rather one of restricting a sufficiently large area around the beam center from normal human activities. According to Glaser (1977), microwave exposure levels beyond a 6.2 mile radius from the beam center would be below the lowest international standards. If the same exclusion area (about 120 square miles) were instead covered with photovoltaic arrays, the amount of energy that could be produced on an annual basis could be comparable to, and possibly greater than, that produced by a 5,000 MWe satellite station.

Another troublesome problem associated with microwave power transmission would be the interference with radio and communication systems. This problem might prove to be not only an annoyance, but possibly hazardous if the normal functioning of navigation and radio communication systems of aircraft were impaired even temporarily.

Finally, SSPS should be carefully assessed with regard to net energy and resource requirements. A recent analysis indicates a net energy ratio that is significantly lower than that of present electrical generation technologies (Herendeen, et al. 1979). In addition to the large liquid fuel demands for the launch vehicles, requirements for critical materials, such as platinum, should be evaluated.

Overall Assessment

In short, the SSPS concept is a risky and expensive proposition, shrouded with environmental uncertainties and offering no compelling advantages over land-based solar options. Therefore, while further research may be justifiable to resolve some of the many outstanding questions, it would be premature in the extreme to launch a major development program leading towards a full scale demonstration.

OCEAN THERMAL ENERGY CONVERSION (OTEC)

Introduction

The oceans, occupying 70 percent of the global surface area, act as vast natural collectors and storage reservoirs for solar energy. Heating by the sun produces temperature differences between the warmer ocean surface waters and the cooler, deeper layers 2,000 to 3,000 feet below the surface. These temperature differences are maintained by the large-scale circulation of the ocean's waters, driven by the sun, and attain their largest and most stable values of about 30° to 45°F in tropical waters within 20° latitude of the equator. Ocean thermal energy conversion systems seek to derive useful energy from these thermal gradients.

The principle attraction offered by OTEC is the fact that the ocean itself both collects and stores solar energy, obviating the need for these two components, generally essential to solar energy systems. Furthermore, the storage capabilities of the ocean allow for continuous plant operation, in theory avoiding the problems of an intermittent energy source inherent to land-based solar energy systems. However, the technical obstacles to ocean thermal conversion are formidable, and consequently, OTEC is not expected to make a major contribution to the nation's energy supply in this century.

Resource Potential

Although the amount of energy stored in the ocean is phenomenally large, only a very small fraction could conceivably be extracted. The energy potential of ocean thermal conversion may ultimately be constrained by the degree of environmental and climatic modification deemed tolerable. Another major limitation on the overall potential for OTEC is the availability of suitable sites with the required high temperature gradients, minimal storm activities, manageably low ocean currents, and proximity to energy use centers. The only serious possibilities for the siting of OTEC plants off the continental U.S. coast are the Florida Gulf Stream and the Gulf of Mexico. Other, more remote sites, are possible in the Caribbean and the Pacific, but have not been considered here because of problems relating to long distance energy transmission.

The Gulf Stream would be a potentially attractive site because the flow of useful energy through a given cross-sectional area is considerably more concentrated than the incident solar flux on the earth's surface (von Hippel and Williams 1975). The local and global impacts of large-scale energy extraction from the Gulf Stream are largely unknown at this time, but the climate of Western Europe could well be affected in some manner if large-scale energy extraction was implemented. Von Hippel and Williams (1975) have estimated that approximately 2 trillion kilowatt hours of electricity (or about 20 quads on a primary energy equivalent basis) per year could be extracted from the Gulf Stream if the warmer surface waters were allowed to undergo a temperature drop of about 0.5°F. Further scientific studies are required to determine whether this is an acceptable upper limit. Other investigators (Gross et al. 1975) have obtained resource estimates of the same order of magnitude.

The energy potential of the relatively still waters of the Gulf of Mexico is much more limited, because considerably greater surface areas would need to be utilized in order to hold surface temperature drops within acceptable limits (von Hippel and Williams 1975). One study found that OTEC plants would have to "graze" more than a 200,000 square mile section of the Gulf in order to produce 1 quad per year (OTA 1978). Thus, this region does not appear to offer a practical site for OTEC systems.

Technology

Current designs employ a cycle in which warm surface water is used to evaporate ammonia. The evaporated ammonia gas is used to drive a turbine generator and is then condensed by cooler waters drawn from the ocean depths. In August of 1979, a small OTEC plant off the coast of Hawaii began to generate 50 kWe of power, 40 kWe of which were used to operate the plant (White 1979). However, many engineering problems need to be resolved before the feasibility of large-scale OTEC can be firmly established.

The fundamental problem confronting OTEC is that the available temperature differentials are so low—less than 45°F—that even an ideal heat engine would have an efficiency below 8 percent. In practice, however, a working OTEC plant is not likely to achieve an efficiency of better than 2 to 3 percent (Metz 1977d). Such low performance necessitates moving immense volumes of water. A 100 MWe OTEC plant would require 30,000 cubic feet of water per second—a flow rate more than two and a half times that of the Potomac River at Washington, D.C. (OTA 1978). The amount of energy required for pumping would be significant—amounting to almost one-third of the plant's net output (Metz 1977d). Suboptimal performance from any of a number of key components of an OTEC plant might jeopardize the production of net energy.

Although no theoretical breakthroughs are required, major engineering advances will be needed to build an OTEC plant because

Photograph 4-13: Ocean Thermal Energy Conversion. Artist's conception of a vertical, semi-submerged, spar-shaped OTEC plant, designed by Lockheed Space and Missile Company. The plant would generate about 250 megawatts of electric power. Its pipes would extend to depths of 1,500 to 3,000 feet into the ocean.

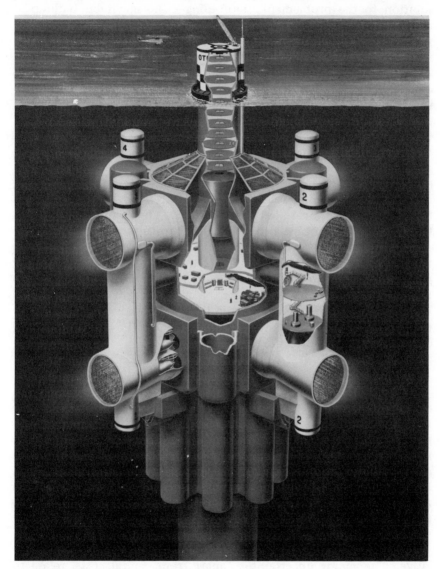

Source: Lockheed Space and Missile Company, Inc.

many of the major components have yet to be developed. The design and construction of the enormous heat exchangers required for OTEC plants poses a major engineering challenge. The most suitable material for this component has yet to be identified. A practical solution to the problem of "biofouling"—the inevitable buildup of marine growths on heat exchangers and other equipment—must be found. Otherwise, the reduced heat transfer would lower the already modest plant efficiency and further degrade the net energy balance. The design and building of a 2,000 or 3,000 foot pipe to draw so much cold water from the lower ocean represents another technological challenge. The deployment of OTEC systems is also dependent upon technological advances in underwater electricity transmission beyond the present state of the art.

Also in need of resolution is the design of an adequate mooring system. Present experience with the mooring problems likely to be encountered is limited. The feasibility of mooring an OTEC plant, or anything of comparable size, in the currents of the Gulf Stream remains to be demonstrated (Metz 1977d). The resultant stresses on the huge cold water pipe also need closer scrutiny.

A frequently cited attribute of OTEC is that it could be applied to the utilization of waste heat discharged to water bodies from power plants, where the temperature gradients would be comparable to those found in the ocean. Perhaps the most prudent approach would be to first test OTEC technology in this land-based application, prior to the construction of prototype models under the more demanding conditions existing offshore (Metz 1977d).

Economics

Capital cost estimates for OTEC plants vary widely, ranging from $500 to $3,700/kWe, with $2,500/kWe considered a reasonable figure (OTA 1978; ERDA 1974; Metz 1977d). However, these estimates rely on an incomplete data base, since many components have not yet been built. Systems built for marine environments are generally much more expensive than their land-based counterparts. For example, technical difficulties resulting in cost overruns by a factor of two to two and a half times the original estimates are typical for offshore oil-drilling plants, which are in some ways similar to OTEC plants (Metz 1977d).

The ultimate cost of OTEC-generated electricity is highly speculative. The major uncertainties, apart from capital costs, relate to plant reliability and lifetime. The reliability of OTEC plants cannot now be predicted and needs to be more closely assessed. Although present concepts assume a capacity factor of 90 to 94 percent—far better than the average performance of land-based plants—available evidence suggests that an upper limit of 80 to 85 percent is more plausible (OTA 1978), and even that figure might be overly optimistic. Unsatisfactory reliability or a limited useful lifetime could prevent OTEC plants from recouping their large capital investments.

Environmental Considerations

There are a number of environmental impacts that might be expected from a network of OTEC plants, none of which has been thoroughly investigated. These impacts vary in scale and severity and might be loosely separated into local and global effects.

Local effects might include the release to the ocean of toxic chemicals, working fluids, and metallic ions from corrosion. Marine life in the vicinity of the plant could also be adversely affected by these as well as by attendant changes in temperature and pressure. Because large volumes of ocean water will be circulated through the plants' heat exchangers, the entrainment of marine organisms might prove troublesome. The upwelling of nutrient- and carbon-enriched water from the ocean depths could also pose serious problems. Calculations performed by Williams (1975) suggest that OTEC plants could release substantial amounts of carbon dioxide to the atmosphere—perhaps one-third as much per unit of electrical output as a fossil fuel power plant.

A reduction of average ocean surface temperatures and altered oceanflow patterns caused by the extensive operation of OTEC plants could have a significant impact on global weather and climate. For example, Northern Europe is much warmer than many other regions at the same latitude (e.g., London is at the same latitude as the much colder Hudson Bay) largely owing to the warm ocean waters carried by the Gulf Stream (Schneider and Mesirow 1976). Therefore, modifications of the Gulf Stream could have a profound influence over the climate of Northern Europe. Certainly any decision regarding large-scale deployment would have to be conditional upon resolution of this question.

Overall Assessment

OTEC is presently an undemonstrated technology. Numerous fundamental technical, economic, and environmental questions need to be resolved before the ultimate role of OTEC in the nation's energy future can be delineated. These questions have to be thoroughly investigated in an orderly research and development program prior to the construction of a large-scale demonstration plant. The initiation of a demonstration program would be costly and premature at the present time. A thorough research and development effort could, within ten years, provide answers to many outstanding questions pertaining to the technical and economic feasibility and environmental acceptability of large-scale power generation from OTEC.

In any event, it is clear that due to environmental and resource constraints, OTEC will at best play a modest and supplementary role in comparison to land-based solar-electric options. Therefore, a rational government funding policy should reflect the relatively small overall contribution that can reasonably be expected from OTEC. If the many remaining problems are satisfactorily resolved, however, OTEC may ultimately provide an important means of baseload power generation for certain regions of the country, but probably not until the next century.

OTHER OCEAN ENERGY RESOURCES

In addition to ocean thermal gradients, energy is also stored and dissipated in the ocean by such phenomena as tides, waves, currents, and salinity gradients. These potential energy resources are reviewed in this section. Although technically feasible, tidal power is constrained by the limited availability of suitable sites. The energy potential of waves could be of significance on a regional basis, but the premature state of conversion technology and economic uncertainties leave open, for now, the question of the ultimate contribution to be expected from this resource. Ocean currents do not appear to be a potentially significant energy source due to the limited resource base and practical difficulties involved in energy conversion. Finally, the utilization of salinity gradients for large-scale electricity generation will be unacceptable on environmental grounds, owing to the consumption of massive amounts of fresh water.

Tides

Tidal energy is not sustained by solar radiation, but rather by gravitational energy in the earth-moon-sun system. The principle behind the conversion of tidal motions into electricity is fairly simple. A dam is constructed in a bay or estuary, creating a basin that is separated from the sea. The dam is used to create a difference in water level between the basin and sea, and the incoming and outgoing movement of the tides is used to drive hydraulic turbines propelling electric generators. More complex systems may use several ponds with flows between them as well as into the ocean to yield steadier power.

Only two tidal power generation plants have been built. The

world's first major tidal electric plant, having a capacity of 240 MWe, was completed at La Rance, France, in 1966 (Gibrat 1966). A small (400 kWe) experimental tidal power plant was also built in the Soviet Union in 1968.

Worldwide, the total tidal power potential is quite small. According to one estimate (Hubbert 1969), only about 13,000 MWe of continuous electrical power is available, equivalent to about 114 billion kWhe annually or 5 percent of total U.S. electricity consumption in 1978. Similarly, Merriam (1978) has estimated the worldwide tidal energy potential to be about 1.3 trillion kWh or 13 quads (primary energy equivalent) per year, only 10 to 25 percent of which could be practically recovered in the form of electricity. The harnessing of tidal power requires a favorable site, capable of impounding large volumes of water and regularly experiencing large variations in water surface levels. The only potential sites available to the United States are the Passamaquoddy Bay, lying on the Maine-Canada border, and the Gulf of Alaska, together representing less than 5 percent of the world's total tidal power potential (Merriam 1978). Thus, tidal power cannot be considered a resource of major significance for the United States, although it could be used as a supplementary power source in the regions specifically mentioned above.

The capital costs of tidal plants appear to be rather high. The cost of the Rance power station was about $530/kWe in 1966 dollars, but would cost much more to construct today. The construction of a 500 MWe plant in the Passamaquoddy Bay of Maine is projected to cost about $1 billion or, equivalently, $2,000/kWe.

On balance, tidal power conversion appears to be relatively benign from an environmental standpoint. The operation of tidal power stations produces no noxious wastes and consumes no depletable energy resources. The experience at La Rance has demonstrated no major adverse impacts on the environment (Waller 1972). One source has claimed, however, that damming the Bay of Fundy in Nova Scotia for tidal power generation could lead to flooding and erosion along the northeastern U.S. coast; possible modifications in local weather patterns (i.e., hotter summers and colder winters) resulting from the creation of a still basin of water were also cited (*Environment* 1978). Before proceeding with the construction of future tidal power facilities, potential effects on the local marine ecology and fishing industries should be carefully examined.

A quite different approach to harnessing tidal power has been proposed involving the use of underwater turbines called "rivermills" (*New Scientist* 1979). In contrast to conventional systems, no dam would be required; instead, vertical axis rotors would be installed in the ocean in regions endowed with strong tidal streams and currents.

Tidal streams can have high power densities greatly exceeding those for wind. At a favorable site, a tidal current might have a power density of about 400 W/ft^2 (4.4 kW/m^2), whereas the corresponding value for a particularly windy region would be not much more than about 50 W/ft^2 (500 W/m^2). The tidal stream power potential for the United States has not been explored, although a total potential of about 15,000 MW has been identified in ocean regions near the United Kingdom. From a technical point of view, the concept is quite immature: some limited small-scale testing has been conducted, but no large-scale demonstration models have been built. An estimated $10 million cost (i.e., $1,000/kWe) for a 10 MWe "rivermill" with a 330 foot (100 m) diameter rotor must be considered highly preliminary in light of the existing uncertainties. If proved feasible, this concept would appear to have substantial environmental advantages over "conventional" tidal power systems given that the structure would be completely underwater and thus unobtrusive. Nonetheless, interference with shipping and fishing could prove to be a problem in congested areas.

In conclusion, although tidal power is a renewable resource, the ultimate energy potential for conventional installations is severely limited by the small number of suitable locations. While small regional contributions are possible, detailed cost comparisons with other renewable energy systems should be made prior to a decision to construct tidal installations. Harnessing the power of tidal streams with underwater turbines presents an option that may hold advantages over the more conventional approach. However, the energy potential of tidal streams is virtually unknown for the United States. Furthermore, the practical feasibility of the requisite conversion technology remains to be demonstrated.

Wave Energy

Wave energy is created by the winds passing across open waters. It involves the gravitational potential energy of elevated water and the kinetic energy from the forward motion of the waves. Schemes to convert wave energy to a usable form vary depending upon the type of motion they are designed to harness. Estimates of the power available from the waves off the U.S. coastline are presently uncertain, but all indicate a maximum energy potential that is relatively small, although not insignificant, in comparison to total U.S. energy use.

The most attractive sites appear to be off the coast of northern California, Oregon, Washington, and Alaska, although a sizeable amount of energy is potentially available from the East Coast as well. Data from Leishman and Scobie (1976) place the annual average

Photograph 4-14: Wave Energy Extraction. Artist's rendering of the Dam-Atoll, a 250 foot diameter generating turbine that would use wave energy to produce a power output of one to two megawatts. The concept was developed at the Lockheed-California Company.

Source: Lockheed-California Company.

wave power off the West and East coasts at about 75 MW/mile and 62 MW/mile, respectively. If a single line of wave energy generators spanning a length of 300 miles were installed off the Northwest coast, assuming continuous operation and an electrical conversion efficiency of 50 percent, approximately 100 billion kilowatt hours (electrical) or about 1 quad per year (primary energy equivalent) could be provided. The possibility also exists for installing multiple rows of wave generators spaced in such a way that the wave energy is regenerated by the wind between successive rows (McCormick 1976). A distance of about thirty miles between parallel rows of wave convertors has been suggested as providing sufficient spacing (Craig et al. 1978). Such a configuration could expand the potential wave energy resource and also allows for the possibility of minimizing the stretch of coastline required to produce a given amount of energy.

In other countries, wave power may have a much greater potential.

It has been estimated that wave power could supply up to one-half of the current electricity requirements of the United Kingdom (Leishman and Scobie 1976) and, conceivably, all of Japan's energy needs (*Mechanical Engineering* 1976).

Although the possibility of obtaining useful energy from waves has been considered for several hundred years, no large-scale conversion systems have yet been built. A number of conversion devices, incorporating floats, buoys, pipes, pumps, and vanes, have been proposed to harness wave energy, but few have ever been operated. Many of the various schemes have been reviewed by Leishman and Scobie (1976) and ERDA (1976). Estimated conversion efficiencies range from 10 to 80 percent (Craig et al. 1978). In Japan, small-scale tests operated on one system actually demonstrated efficiencies up to 80 percent (*Mechanical Engineering* 1976). At present, wave energy conversion is in the research, design, and experimentation stages. No demonstrated technologies are presently available. Numerous problems, such as biofouling, corrosion, and mechanical breakdown, await definitive resolution as the technology matures.

Cost estimates for wave power systems are necessarily uncertain and range from about $850/kWe (Salter et al. 1976) up to $2,400/kWe, including transmission equipment (Leishman and Scobie 1976). Further research and development will be required to determine the cost of wave-generated electricity more accurately.

The environmental impacts associated with wave power systems do not appear to be severe. No chemical pollutants or heat will be added to the biosphere as a result of energy extraction from waves, although a minute cooling of ocean water would result. In general, wave power stations could be sited with a low profile or far enough offshore so as to not affect the visual environment. Benefits may possibly be derived from using the floating generators as breakwaters for fish breeding grounds. Extensive arrays of wave convertors would pose obstacles, and thus hazards, to marine transport (Probert and Mitchell 1979). The potential severity of this problem needs to be more fully explored. Although no other major impacts are readily apparent at present, further investigation of potential effects on coastal erosion and other possible problems should be investigated prior to widespread deployment of these devices.

Although the energy potential of the wave resource is limited, power provided from waves could certainly be important on a regional basis. However, the future deployment of wave power systems is contingent upon the resolution of numerous technical and economic uncertainties.

Ocean Currents

The generation of electricity from the energy contained in ocean currents has also been proposed (Stewart 1974). The energy potential of this option appears to be very limited. The most favorable sites investigated to date are located off the east coast of Florida. The power contained in the currents of the Florida Gulf Stream has been estimated to be about 25,000 MW (MacArthur 1974; *New Scientist* 1979), but only a minute fraction could conceivably be extracted in order to prevent "stopping" or dramatically altering the Gulf Stream. In addition, practical power generation awaits the development of suitable technology. The underwater rotor (or "river-mill") mentioned previously in this chapter offers one possible approach to tapping the energy of the Florida current (*New Scientist* 1979). The technical and economic feasibility of this concept, however, remains to be verified. In any event, energy from ocean currents clearly does not represent a significant resource for the United States.

Salinity Gradients

Salinity gradients are indirectly produced and maintained by solar energy: the sun constantly evaporates ocean water, making it more saline and simultaneously producing fresh water that, when rained on land, returns to the sea in rivers. A potential source of energy lies at the interface between fresh and salt waters, such as areas where rivers flow into the ocean. The difference in salinity between the fresh and salt water can establish a pressure differential across a semipermeable membrane, equivalent to the pressure of a water column approximately 800 feet high (Craig et al. 1978). Methods proposed for tapping this osmotic pressure difference to yield useful power involve either utilizing the pressure differential to drive a turbine or utilizing the electric potential created between solutions having different salt concentrations (ERDA 1976; Weinstein and Leitz 1976; Clampitt and Kiviat 1976). In theory, the mixing of about 264 gallons (1 cubic meter) per second of fresh water with sea water could release 1 megawatt of power, which conceivably could be converted to electricity with an efficiency of 15 to 50 percent (Craig et al. 1978). However, utilization of salinity gradients will be hampered by its unproven economic and technological feasibility, serious environmental problems, and constraints on the availability of fresh water.

No demonstrated technology exists today for the large-scale generation of power from salinity gradients. Much research and development work must be done before a practical system can be made available. Furthermore, major reductions in the costs of components, particularly in the manufacture of suitable membranes, are required in order to place salinity gradient power conversion upon a competitive basis with other energy sources (Wick and Isaacs 1976).

Environmental problems pose the major barrier to the development of this resource. Large-scale exploitation is likely to have a disastrous effect on estuaries (Ehrlich Ehrlich, and Holdren 1977). The dominant problem, however, relates to water use. Salinity gradient power plants would require enormous amounts of fresh water. In one year, assuming a generous 50 percent efficiency, a continuously operating 1,000 MWe plant would consume about 17 trillion gallons of fresh water, amounting to about 4 percent of the total annual river runoff in the United States. Producing 1 quad per year from salinity gradient power would require roughly 40 percent of the total river runoff.

In conclusion, the early state of technological development and the presently unfavorable economics both cast doubt on the future development of salinity gradient power. By far the preeminent problem, however, is the excessive demand for fresh water, which alone would render salinity gradient power unacceptable except on an extremely limited basis.

*   *   *   *

We have thus reviewed the major renewable energy resources and technologies that are either available or proposed for harnessing them. As we have seen, some of these options look very promising while others appear marginal at best. In addition to being able to collect energy from renewable sources in the proper quantity and form, however, it is also necessary to provide some means of storing and transporting energy in order to be able to supply it when and where it is needed. Energy storage and transmission are therefore the subjects of the next chapter. Finally, in Chapter 6, a solar energy strategy capable of meeting future U.S. energy needs is outlined on the basis of material presented in the two preceding chapters. Among all the renewable energy systems reviewed here, the most attractive are selected and matched with appropriate energy-requiring tasks in the U.S. economy.

REFERENCES

Allen, A.E. 1976. "Potential for Conventional and Underground Pumped-Storage." Paper presented at IEEE/ASME/ASCE Joint Power Generation Conference at Buffalo, New York, September 19–23.

American Institute of Aeronautics and Astronautics (AIAA). 1975. "Solar Energy For Earth: An AIAA Assessment." New York.

American Physical Society (APS). 1979. "Executive Summary of the American Physical Society Study Group on Solar Photovoltaic Energy Conversion." New York, January.

Anderson, B.A., and C.J. Michal. 1978. "Passive Solar Design." *Annual Review of Energy* 3:57-100.

Antal, M.J. 1976. "Tower Power: Producing Fuels From Solar Energy." *Bulletin of the Atomic Scientists* 32, no. 5 (May):59-62.

———. 1979. Princeton University, telephone conversation, January 6.

Army Corps of Engineers. 1975. "Summary of Northwest Hydroelectric Power Potential." Washington, D.C., February. rev.

———. 1977. "Estimate of National Hydroelectric Potential at Existing Dams." Washington, D.C.: Institute of Water Resources, July.

Audubon Magazine. 1975. 77, no. 6 (November): 137.

———. 1976. "Econotes." 78, no. 2 (March):140.

Blackwell, B.F., and L.U. Feltz. 1975. "Wind Energy—A Revitalized Pursuit." SAND 75-0166. Albuquerque: Sandia Laboratories, March.

Brodeur, P. 1976. "Microwaves." Part I and Part II. *New Yorker*, December 13 and 20.

Caputo, R.S. 1977. "Solar Power Plants: Dark Horse in the Energy Stable." *Bulletin of the Atomic Scientists* 33, no. 5 (May):47-56.

Carpenter, P.R., and G.A. Taylor. 1978. "An Economic Analysis of Grid-Converted Residential Solar Photovoltaic Power Systems." Report no. MIT-EL-78-007. Cambridge, Mass.: MIT Energy Laboratory, May.

Chalmers, B. 1976. "The Photovoltaic Generation of Electricity." *Scientific American*, October, pp. 34-43.

Chambers, R.S., et al. "Gasohol: Does It or Doesn't It Produce Positive Net Energy." Urbana, Ill.: Energy Research Group, University of Illinois, September.

Charlton, R.E. 1979. Vice President, Solectro-Thermo, Inc., Dracut, Massachusetts, letter dated January 22 and brochure.

Chemical and Engineering News (C&EN). 1979. "Biomass Potential in 2000 Put at 7 Quads." 57, no. 7 (February 12).

Clampett, B.H., and F.E. Kiviat. 1976. "Energy Recovery From Saline Water by Means of Electrochemical Cells." *Science* 194, (November 12):719-20.

Commoner, B. 1978. "The Solar Solution." *Time and Energy*, November-December, pp. 47-48.

———. 1979. "The Solar Transition—1." *The New Yorker*, April 23.

Corso, R.A. 1976. "Private Sector Hydroelectric Development." Paper presented at IEEE/ASME/ASCE Joint Power Generation Conference at Buffalo, New York, September 19-23.

Council on Environmental Quality (CEQ). 1978. "Solar Energy: Progress and Promise." Washington, D.C., April.

Craig, P., et al. 1978. "Distributed Energy Systems in California's Future." Interim Report, vols. I and II. Prepared for the U.S. Department of Energy. Washington, D.C.: Government Printing Office, May.

Department of Energy (DOE). 1978a. "Budget Highlights." FY 1979 Budget to Congress. January.

———. 1978b. "Status Report on Solar Domestic Policy Review" (public review copy). Washington, D.C., August 25.

———. 1979a. "Modest Impact Seen For Alcohol Fuels Until Conversion Facilities Expand. *Energy Insider*, July 23, p. 1.

———. 1979b. "Monthly Energy Review." Washington, D.C.: Energy Information Administration, DOE/EIA-0035(79), February.

Dietz, J. 1978. "Passive Solar Heat: A First." *Boston Globe*, February 23.

Dionne, E.J., Jr. 1979. "Three State Agencies Battle for Rights to City Garbage." *New York Times*, August 8, p. B5.

Donovan, C., et al. 1979. "Energy Self-Sufficiency in Northhampton, Massachusetts." Amherst, Mass.: Hampshire College, January 3.

Duff, W.S., and W.W. Shaner. 1976. "Solar Thermal Electric Power Systems: Manufacturing Cost Estimation and Systems Optimization." ASME Paper No. 76-WA/HT-14. New York: American Society of Mechanical Engineers.

Eckholm, E.P. 1975. "Salting the Earth." *Environment* 17, no. 7 (October): 9-15.

Ehrlich, P.R.; A.H. Ehrlich; and J.P. Holdren. 1977. *Ecoscience: Population, Resources, Environment.* San Francisco: W.H. Freeman and Company.

Eldridge, F. 1975. "Wind Machines." McLean, Va.: MITRE Corporation, MTR-6971, October.

———. 1976a. "Wind Energy Conversion Systems Using Compressed Air Storage." McLean, Va.: MITRE Corporation, M76-39.

———. 1976b. Telephone conversation, November 3.

Electric Power Research Institute (EPRI). 1976. "An Assessment of Energy Storage Systems Suitable for use by Electric Utilities." EPRI EM-264, ERDA E (11-1)-2501. Final Report, vol. 1. Palo Alto, Calif., July.

Electrical World. 1976. "World News Beat." 135, no. 1 (January 1): 15.

Energy Policy Office. 1976. "The Use of Solar Energy for Space Heating and Hot Water." Boston: Commonwealth of Massachusetts, April.

Energy Research and Development Administration (ERDA). 1975. "Ocean Thermal Energy Conversion Research on an Engineering Evaluation and Test Program." Prepared by TRW Systems Group Global Marine Development, Inc., and United Engineers and Constructors under Contract No. NSF-C958.

———. 1976. "Wave and Salinity Gradient Energy Conversion." ERDA Report No. Coo-3946-1. Workshop Proceedings, May, p. 24-26.

Engelke, C.E. 1978. "A Self-Contained Community Energy System." *Bulletin of the Atomic Scientists*, November, pp. 51-53.

Environment. "Spectrum." 20, no. 9 (November): 23.

Erskine, G.S. 1978. "A Future for Hydropower." *Environment* 20, no. 2 (March): 33-38.

Fan, J.C.C. 1978. "Solar Cells: Plugging Into the Sun." *Technology Review* 80, no. 8 (August-September).

Federal Energy Administration (FEA). 1974. "Project Independence." Final Task Force Report—Solar Energy. Washington, D.C., November.

———. 1974a. "Project Independence Report." Draft copy. Washington, D.C.

———. 1977. "Preliminary Analysis of an Option for the Federal Photovoltaic Utilization Program (FPUP)." Washington, D.C., July 20.

Federal Power Commission (FPC). 1975. "Hydroelectric Plant Construction Cost and Annual Production Expenses: Sixteenth Annual Supplement, 1972." Report FPCS-243. Washington, D.C., January.

Feldman, J.F., et al. 1973. "Cattle Manure to Pipeline Gas." *Mechanical Engineering*, October.

Forest Service. 1973. "Forest Residues—A Resource." Madison, Wisc.: Forest Products Laboratory, Department of Agriculture.

———. 1974. "The Outlook for Timber in the United States." Forest Service Resource Report No. 20. Washington, D.C.: Department of Agriculture, July.

———. 1975. "A Summary of the Program and Assessment for the Nation's Renewable Resources" and "The Nation's Renewable Resources—An Assessment, 1975." Both prepared under the Forest and Rangeland Renewable Resources Planning Act of 1974. Washington, D.C.: Department of Agriculture, August.

Frank, A. 1978. "PV Costs: The Great Debate." *Solar Age* 3, no. 8 (September): 42.

General Electric Space Division. 1977. "Wind Energy Mission Analysis." Prepared for ERDA, No. 765054267. February.

Gibrat, R. 1966. "Scientific Aspects of the Use of Tidal Energy." *Revue Française De L'Energie* 183 (September-October).

Glaser, P.E. 1977. "Solar Power From Satellites." *Physics Today*, February.

Glicksman, L.R. 1978. "Heat Pumps: Off and Running Again." *Technology Review*, June-July, pp. 64–70.

Green, A.W., and T.S. Setzer. 1974. "The Rocky Mountain Timber Situation, 1970." USDA Forest Service Research Bulletin INT-10. Intermountain Forest and Range Experiment Station, November.

Gross, W.P., et al. 1975. "Summary of University of Massachusetts Research on Gulf Stream Based Ocean Thermal Power Plants." Proceedings of the Third Workshop on Thermal Energy Conversion. Houston, Texas. Laurel, Md.: The Johns Hopkins University Applied Physics Laboratory, APL/JHUSR75-2.

Groth, E., III. 1975. "Increasing the Harvest." *Environment* 17, no. 1 (January-February).

Gustavson, M.R. 1979. "Limits to Wind Power Utilization." *Science* 204 (April 6):13–17.

Hammond, A.L. 1977a. "Photovoltaics: The Semiconductor Revolution Comes to Solar." *Science* 197 (July 29):445–47.

———. 1977b. "Photosynthetic Solar Energy: Rediscovering Biomass Fuels." *Science* 197 (August 19):745–46.

Hammond, A., and W.D. Metz. 1978. "Capturing Sunlight: A Revolution in Collector Design." *Science* 201 (July 7):36–39.

Harte, J., and A. Jassby. 1978. "Energy Technology and Natural Environments: The Search for Compatibility." *Annual Review of Energy* 3:101–46.

Herendeen, R.A., et al. 1979. "Energy Analysis of the Solar Power Satellite." *Science* 205 (August 3):451–54.

Herman, S.W., and J.S. Cannon. 1976. *Energy Futures.* New York: Inform, Inc.

———. 1977. *Energy Futures: Industry and the New Technologies.* Cambridge, Mass.: Ballinger Publishing Company.

Heronemus, W.E. 1976. "Oceanic Windpower." Paper presented at the Conference on "Energy from the Oceans: Fact or Fantasy?", North Carolina State University, January.

Hirshorn, A. 1979. "Answer: Energy as an End Product." *New York Times,* August 6, p. A17.

Hubbert, M.K. 1969. "Energy Resources." In *Resources and Man,* ch. 8. San Francisco: W.H. Freeman and Company.

Huetter, J., et al. 1978. "Status, Potential, and Problems of Small Hydroelectric Power Development in the United States." Prepared for the Council on Environmental Quality by Energy Research and Applications, Inc., March.

Inglis, D.R. 1978. "Power from the Ocean Winds." *Environment* 20, no. 8 (October):17–20.

Intertechnology Corporation (ITC). 1977. "An Analysis of the Economic Potential of Solar Thermal Energy to Provide Industrial Process Heat." Warrenton, Va., February.

Justus, C.G. 1976. "Wind Energy Statistics for Large Arrays of Wind Turbine (New England and Central U.S. Regions)." Prepared for the National Science Foundation, ERDA/NSF-00547/76/1, August.

Justus, C.G., et al. 1976. "Nationwide Assessment of Potential Output from Wind-Powered Generators." *Journal of Applied Meteorology* 15, no. 7 (July).

Kasper, W.C. 1974. "Power from Trash." *Environment* 16, no. 2 (March).

Kelly, H. 1978. "Photovoltaic Power Systems: A Tour of the Alternatives." *Science* 199 (February 10):634–43.

Leishman, J.M., and G. Scobie. 1978. "The Development of Wave Power— A Techno-Economic Study." Glasgow: Economic Assessment Unit, National Engineering Laboratory.

Lockheed. 1976. "Wind Energy Mission Analysis." Executive Summary. Prepared for Energy Research and Development Administration, SAN/1075-113. Burbank, Calif.: Lockheed-California Company, October.

Lorentz, E.N. 1967. *The Nature and Theory of the General Circulation of the Atmosphere.* Geneva: World Meteorological Organization.

MacArthur. 1974. Proceedings of the MacArthur Workshop on the Feasibility of Extracting Usable Energy from the Florida Current, Palm Beach Shores, Florida, February 27–March 1.

Maidique, M. 1979. "Solar America." In *Energy Future,* R. Stobaugh and D. Yergin, eds., p. 188. New York: Random House.

Margen, P. 1978. "Central Plant for Annual Heat Storage." *Solar Age,* October, pp. 22–26.

McCormick, M.E. 1976. "Salinity Gradients, Tides, and Waves as Energy

Sources." Paper presented at the Conference on "Energy from the Oceans: Fact or Fantasy?", North Carolina State University, January.

McDowell, E. 1979. "Solarville, Ariz.: Down to $7 a Watt." *New York Times*, June 24.

McQuiston, J.T. 1979. "Gasohol: The Issue is Practicality." *New York Times*, May 19.

Mechanical Engineering. 1976. "Japan Looks at Wave Power Generation." August.

Meinel, A.B., and M. P. Meinel. 1976. *Applied Solar Energy: An Introduction.* Reading, Mass.: Addison-Wesley Publishing Company.

Merriam, M.F. 1977. "Wind Energy for Human Needs." *Technology Review*, January.

———. 1978. "Wind, Waves, and Tide." *Annual Review of Energy* 3:29-56.

Merrigan, J.A. 1975. *Sunlight to Electricity: Prospects for Solar Energy Conversion by Photovoltaics.* Cambridge, Mass.: The MIT Press.

Metz, W.D. 1977a. "Solar Thermal Electricity: Power Tower Dominates Research." *Science* 197 (July 22):353-56.

———. 1977b. "Solar Thermal Energy: Bringing the Pieces Together." *Science* 197 (August 12):650-51.

———. 1977c. "Wind Energy: Large and Small Systems Competing." *Science* 197 (September 2):971-73.

———. 1977d. "Ocean Thermal Energy: The Biggest Gamble in Solar Power." *Science* 198 (October 14):178-80.

Metz, W.D., and A.L. Hammond. 1978. *Solar Energy in America.* Washington, D.C.: American Association for the Advancement of Science.

MITRE Corporation. 1975. "Proceedings of the Second Workshop on Wind Energy Conversion Systems." F. Eldridge, ed. NSF/RA/N-75-050, MTR-6970. McLean, Va., June.

———. 1976. "An Economic Analysis of Solar Water and Space Heating." M76-79. McLean, Va., November.

Mlavsky, A.I. 1976. "Photovoltaics: An Interview with A.I. Mlavsky." *Solar Age* 1, no. 4 (April):18-21.

Nadis, S.J. 1977. "The Economics of Solar Heating and Cooling in Energy-Efficient Buildings." Cambridge, Mass.: Union of Concerned Scientists.

National Research Council. 1976. "Renewable Resources for Industrial Materials." PB-257-357. Washington, D.C.: National Academy of Sciences.

National Science Foundation (NSF/NASA Solar Energy Panel). 1972. "An Assessment of Solar Energy as a National Energy Resource." Washington, D.C., December.

New Scientist. 1977. "Fluorescent Collectors Concentrate the Sun Wonderfully." December, p. 66.

———. 1979. "Windmills Could Come in with the Tide." February 15, p. 481.

New York Times. 1979. "20 California Windmills Planned to Supply Power to 1,000 People." April 22.

Odell, D.J. 1975. "Silt, Cracks, Floods, and other Dam Foolishness." *Audubon* 77, no. 5 (September).

Office of Technology Assessment (OTA). 1978a. "Ocean Thermal Energy Conversion." In *Renewable Energy Resources*, pt. I. Washington, D.C.: Government Printing Office, May.

——. 1978b. "Application of Solar Technology to Today's Energy Needs."

——. 1979. "Gasohol: A Technical Memorandum." Washington, D.C.: Government Printing Office, September.

Vol. 1. Washington, D.C.: Government Printing Office, June.

Pimental, D. et al. 1976. "Land Degradation: Effects on Food and Energy Resource." *Science* 194 (October 8):149-55.

Poole, A.D. 1974. "Biological Energy Sources." Cambridge, January.

——. 1975. "The Potential in Energy Recovery from Organic Wastes." In *The Energy Conservation Papers*. Cambridge, Mass.: Ballinger Publishing Company.

Poole, A.D., and R.H. Williams. 1976. "Flower Power: Prospects for Photosynthetic Energy." *Bulletin of the Atomic Scientists* 32, no. 5 (May):48-58.

Potworowski, J.A., and B. Henry. 1976. "Harnessing the Wind." *Conserver Society Notes*, Fall.

Probert, K., and R. Mitchell. 1979. "Wave Energy and the Environment." *New Scientist*, August 2, pp. 371-73.

Reed, T.B. 1976a. "Efficiencies of Methanol Production from Gas, Coal, Waste, or Wood." Paper presented at the American Chemical Society symposium on Net Energetics of Integrated Synfuel Systems, New York, April 5-6.

——. 1976b. "The Production and Use of Alcohols as Fuels." Paper presented at the American Chemical Society symposium "Shaping the Future of the Rubber Industry," San Francisco, October 5-6.

Reed, T.B., and R.M. Lerner. 1973. "Methanol: A Versatile Fuel for Immediate Use." *Science* 182 (December 28).

Rutty, R.M., et al. 1975. "Net Energy from Nuclear Power." IEA-75-3. Oak Ridge, Tenn.: Institute for Energy Analysis, Oak Ridge National Laboratory, November.

Salter, S.H., et al. 1976. "Wave-Power: Nodding Duck Wave Energy Converters." Paper presented at the Conference on "Energy from the Oceans: Fact or Fantasy?", North Carolina State University, January.

Sarkanen, K.V. 1976. "Renewable Resources for the Production of Fuels and Chemicals." *Science* 191 (February 20).

Schneider, S.H., and L.E. Mesirow. 1976. *The Genesis Strategy.* New York: Plenum Press.

Schuon, M. 1979. "Alcohol: The Fuel of the Future?" *New York Times*, May 22.

Sierra Club Bulletin. 1975. "Federal Report Blasts Timber Industry Practices." 60, no. 7 (August-September):24.

Smil, V. 1977. "Energy Solution in China." *Environment* 19, no. 7 (October): 27-31.

Smith, O.J.M. 1976. "Multimode Practical Solar-Thermal-Electric Power Plants." Proceedings of conference on Frontiers of Power Technology, Oklahoma State University, October 27-28.

Solar Energy Digest (SED). 1977. "Arkla Introduces Production Model of Its Solar Home Air Conditioner." 1, no. 8 (January).

———. 1978a. "New Varian Solar Cells Achieve 28.5% Efficiency." 11, no. 2 (August):8.

———. 1978b. 11, no. 6 (December):5.

Solar Energy Intelligence Report (SEIR). 1978a. "Physicists Claim Advances in Photovoltaics May Lead to Low Cost Cells." 4, no. 15 (April 10):101.

———. 1978b. "TEAM Energy PV Projects Underway with Costs as Low as $1.80/pW." 4, no. 20 (May 15):140.

———. 1978c. "Westinghouse Dendritic Web Material Achieves 15.5% Conversion Efficiency." 4, no. 43 (October 23):326.

———. 1979a. "Monsanto Says Miamisburg Solar Pond System Working Well." 5, no. 31 (July 30):308.

———. 1979b. "Large Windmill to be Linked to Hydro at Medicine Bow, Wyoming by Interior, NASA." 5, no. 21 (May 21): 206.

———. 1979c. "California Agency Plans to Buy Electricity from Windmill Grid." 5, no. 17 (April 23):164.

Solarwork. 1979. "Most-Asked Questions about the Solar Industry and Jobs." Sacramento, Calif.: State of California Governor's Office of Appropriate Technology, July, p. 2.

Sorenson, B. 1976. "Dependability of Wind Energy Generators with Short Term Energy Storage." *Science* 194, pp. 935-37.

Spaulding, A. 1979. WTG Energy Systems, Inc., Angola, New York, telephone conversation, February 12.

Stanford Research Institute (SRI). 1977. "Solar Energy in America's Future—A Preliminary Assessment." 2nd ed. Prepared for the Energy Research and Development Administration. DSE-115/2. Menlo Park, Calif., March.

Stewart, H.B. 1974. "Current from the Current." *Oceanus* 17 (Summer).

Taylor, T.B. 1977. "The Ultimate Source." *Skeptic* 18 (March-April).

———. 1978. "In Conversation with Ted Taylor." *Solar Age*, August, pp. 22-25.

Vant-Hull, L.L., and A.F. Hildebrandt. 1976. "Solar Thermal Power Based on Optical Transmission." *Solar Energy* 18:31-39.

Von Hippel, F., and R.H. Williams. 1975. "Solar Technologies." *Bulletin of the Atomic Scientists* 31, no. 9 (November):25-31.

Waller, D.H. 1972. "Environmental Effects of Tidal Power Development." In *Tidal Power.* New York: Plenum Press.

Weinstein, J.N., and F.B. Leitz. 1976. "Electric Power from Differences in Salinity: The Dialytic Battery." *Science* 191 (February 13):557-59.

Wetmore, W.C. 1976. "Vertical-Vortex Wind Turbine Proposed." *Aviation Week and Space Technology*, March 1.

White, H. 1979. "Mini-OTEC." *Solar Energy Digest* 13, no. 3 (September):1.

Wick, G.L., and J.D. Isaacs. 1976. "Utilization of the Energy from Salinity Gradients." La Jolla, Calif.: Scripps Institution of Oceanography.

Williams, J.R. 1974. *Solar Energy—Technology and Applications.* Ann Arbor, Mich.: Ann Arbor Science.

Williams, R.H. 1975. "The Greenhouse Effect and Ocean-Based Solar Energy Systems." Princeton, N.J.: Center for Environmental Studies, Princeton University, October. Working Paper no. 21.

Winston, R. 1976. "A Funnel for the Sun." *Solar Age*, February, pp. 22-25.

Witkin, R. 1979. "Satellites Studied as Power Source." *New York Times*, February 20.

Yen, J. 1976. "Tornado-Type Wind Energy Systems: Basic Considerations." International symposium on Wind Energy Systems, St. John's College, Cambridge, England, September 7-9.

✳ *Chapter 5*

Energy Storage and Transmission

INTRODUCTION

Energy use in the major sectors of the economy can vary considerably with the time of day, from one day to another, and from one season to the next. Energy supplies may also be acquired or generated at rates that can vary appreciably, especially when one considers intermittent sources, such as direct solar energy, whose availability is strictly limited to daylight hours and is often affected by the weather. Adjusting variable energy supplies to variable levels of energy use is a subtle and vital systems problem that needs to be resolved for the establishment of an overall energy supply network. It is most important in integrating intermittent sources into the conventional energy supply system.

The development of satisfactory methods for storing energy provides the key to assuring that sufficient quantities of energy are made available when needed. An energy storage system is quite simply a buffer device, accumulating surplus energy supplies when the rate of production exceeds the level of demand and providing supplemental energy when production falls below consumption. Such an "energy bank," to put it prosaically (but aptly for the case of direct solar power), is a way of saving energy "for a rainy day." An extensive short- and intermediate-term storage system is already available for conventional energy sources (e.g., oil storage depots, pumped hydroelectric facilities, home oil tanks, and the fuel tanks in transportation vehicles), but the development of advanced storage systems suitable for use with renewable energy sources such as solar power remains a major challenge for the future.

In addition to energy storage, there are the related problems of energy transmission, distribution, and conversion. Energy is required at certain locations in specific forms but may be available only at other locations and in other forms. How to transport energy from the point of production to the point of end use and make it available in a suitable form is a critical element in the design of a national energy system. Questions of whether to store energy at the point of production, at the point of use, or at some intermediary site will strongly influence the structure of an overall energy distribution network.

The topics of energy storage and transmission are frequently neglected in the current literature. We have given these issues special attention in this analysis because of their great importance in demonstrating the feasibility of an energy system based on renewable energy sources. Energy storage and transmission are the focus of this chapter. In the first section, the need for energy storage is discussed, and the major storage options are assessed. The role of energy storage in both renewable and conventional energy systems is examined first. The transmission of energy in the forms of electricity and hydrogen is then treated in the following two sections. Strictly speaking, neither electricity nor hydrogen are energy sources or primary energy forms. Rather, they are energy carriers or intermediate forms of energy convenient for purposes of distribution, storage, and utilization. They must be made from primary fuels or primary energy sources. Both of these energy carriers can play critical roles in a solar-powered economy. The relative utility, merits, and disadvantages of electricity and hydrogen are carefully evaluated.

ENERGY STORAGE

Introduction

As a means of balancing varying rates of production and consumption of many materials, storage is an integral part of the economy. For example, the economy is highly dependent on the storage of food, water, fuels, and other resources. Businesses accumulate massive inventories of key materials and intermediate and finished products. There is in fact quite a science behind commercial inventory practices used in the determination of optimal inventory sizes and in weighing the costs and benefits of storage.

Not surprisingly, energy storage (broadly defined) is not a novel concept, but has in fact been utilized in various forms throughout history. For example, all societies have relied to some extent on the solar energy captured by green plants through photosynthesis. Many civilizations have constructed buildings out of bulky, massive materials such as mud and adobe that are capable of storing solar energy and thereby providing for relatively uniform temperature environments. In more recent times, rivers have been dammed, with the water stored for meeting future power demands.

At the present time, most of the energy used by modern industrial societies, including the United States, is "stored" in the form of fossil fuels. These fuels, formed from partially decayed plants over periods of hundreds of millions of years, indirectly store energy originally derived from sunlight. In addition to the natural storage of fuel resources in underground deposits, artificial surface storage arrangements such as oil tanks have been introduced as well. While fossil fuels were cheap and readily available, alternative storage approaches were given little consideration. However, the imminent depletion of inexpensive supplies of these fuel resources necessitates the development and ultimate deployment of new storage technologies.

Because of the intermittent nature of solar and wind energy, storage will be an essential feature of a renewable energy economy. For example, thermal energy will have to be stored for building heating and for industrial process heat applications. A variety of techniques will be required to store energy to produce electricity in order to balance variations in output from solar and wind generators against a variable demand. Electrical storage batteries capable of propelling electric vehicles may play a particularly significant role in the transportation sector. Energy storage can also be very useful in conjunction with conventional electric power plants and would become a virtual necessity in a large-scale nuclear power program by accommodating the fluctuations in electric demand now handled by oil- and gas-fired intermediate and peak load power plants. Thus, energy storage will become necessary in the long term and can be beneficial in the short run as well.

As an integral component of a rational energy strategy, storage will require a diversified approach. Energy will need to be stored in various forms—thermal, chemical, mechanical, and electrical—by techniques that are compatible with the energy source and the end use. Storage will be required on a variety of scales to serve a broad range of stationary and mobile applications. Some storage methods, such as pumped hydroelectric storage, will only be economically feasible on a large scale in conjunction with centralized power plants, often located far from the point of end use. Other storage systems of a more modular nature, such as thermal stores, batteries, and flywheels, are suitable for decentralized, on-site applications and can be readily matched to diverse loads.

Siting storage units near their loads may reduce energy demand peaks on electric transmission lines. This practice could lead to savings in the cost of transmission and distribution, which frequently account for roughly one-half of a user's electricity bill (Baughman and Bottaro 1976). These expenses are in part attributable to the fact that power lines and distribution equipment may be sized to handle "line peaks," but are generally used at only a small fraction of their capacity. By minimizing load peaks on transmission lines, line losses would be correspondingly reduced (NAS 1976). Another modest advantage of on-site storage would be a reduced vulnerability to temporary transmission and distribution line outages. Before deciding between centralized or on-site storage, however, the potential transmission and distribution savings must be weighed against the cost of power-conditioning equipment required to convert from alternating current to direct current and vice versa.

The choice of an appropriate storage system will also depend upon the duration of the required operation cycle. In some peaking power applications, only short-term (on the order of a few hours) storage may be needed. A storage system operating on a daily charge-discharge cycle in coincidence with the diurnal variations of the sun may be appropriate in some cases. In other cases, longer term storage may be required to take into account seasonal variations in the solar and/or wind resource. Seasonal storage, which would allow for the utilization of excess solar energy collected during the summer, may be especially appealing for solar space heating and power generation applications.

Because energies from the sun and winds fluctuate in time, most practical applications will ultimately require storage to provide energy upon demand. In the near term, however, solar and wind energy systems can operate without storage by relying on auxiliary sources, such as fossil fuels, to provide backup power. In this "fuel saver" mode of operation, any energy supplied by the solar and wind plants will simply save fuel, but not displace capacity. For example, an oil-burning furnace could be used to back up a solar heating system. Similarly, solar and wind-generated electricity could be fed directly into an electrical utility grid, supplying power that would otherwise have come from conventional plants. Under this arrangement, the benefits of solar utilization and attendant fuel conservation would have to be weighed against the costs of maintaining intermittently idle conventional generating capacity. This approach would nonetheless be technically feasible until the fraction of power provided by intermittent sources reaches a level of about 10 to 20 percent (Metz 1978; APS 1979). Beyond this point, however, the introduction of additional solar generation facilities without storage (and hence variable in output) would undermine the overall stability of the electrical system. However, given the characteristically slow turnover times in the electricity-generating sector, the introduction solar-electric technologies will occur gradually throughout the remainder of this century. Therefore, the need for storage should not pose any major barrier to the near-term utilization of solar energy. The time required for solar penetration to reach critically high levels will be more than sufficient to allow for the development and deployment of advanced storage systems.

In the long run, however, energy storage will ultimately assume a significance roughly comparable to that of energy production. Therefore, a commensurate developmental effort should be made during the transitional period. To neglect storage during this time of relatively relaxed sense of urgency would be a profound misjudgment that could seriously delay the establishment of a renewable energy economy.

Energy storage could also play an important near-term role in conjunction with conventional electricity systems. Utility companies have traditionally met the fluctuating demand for electric power by a "generation mix" made up of three components: baseload operating plants—to provide a steady power output; intermediate generating units—to cover the broad peaks in daily demand; and peaking units—to meet the peak power demands.[1] Although this mix has worked adequately in the past, the relatively low operating efficiencies of the intermediate and peaking units, coupled with highly escalated fuel costs, make this approach increasingly expensive and wasteful of diminishing fossil fuel reserves. In addition, the sharp rise in the capital costs of nuclear and fossil baseload plants makes a high capacity factor an economic necessity.

Thus, it would appear to be advantageous from both an economic and energy efficiency point of view to combine baseload plants, operating at full capacity where they are most efficient, with an energy storage system to provide peaking and emergency power requirements, provided such a system is economically competitive. One

---

1. Baseload generation normally provides 40–50 percent of the total generating capacity and about 75 percent of the total annual energy output. Intermediate and peaking plants normally provide the remaining 50–60 percent of the power demand and about 25 percent of the annual electric energy (APL 1975).

possible arrangement would be to continuously operate baseload plants, storing any excess energy in the form of high temperature steam. During periods of high demand, the stored steam could be passed through a turbine to provide electricity. The only alternative to this "load leveling" or "peak-shaving" approach would be to build enough power plants to meet the peak demand and to expect some of them to lie idle a large fraction of the time. Therefore, energy storage can defer the need to construct new intermediate and peak load generating capacity and can also save oil and gas—the fuels upon which these plants usually run. Energy can be saved by eliminating some fraction of the idling combustion turbines now constituting the "spinning reserve." These generators are kept fired up in the event that a large generating unit goes down, but do not normally feed power into the grid. One study concluded that energy storage systems offered electric utilities an attractive alternative to expansions in intermediate and peak load generation equipment that could practically and economically displace 10 percent of the total generating capacity considered necessary for 1985–1990 (NAS 1976). Another study indicated that the savings from thermal storage in utilities would exceed storage costs by a factor of two to four (Asbury and Muller 1977).

In a long-range solar energy future, a broad mix of storage technologies—compatible with the form of energy to be collected and utilized—will likely be employed. Thus, for solar space conditioning, low temperature thermal energy storage is the most logical approach. In solar thermal power plants, high temperature thermal storage is preferable to electrical energy storage, both on the basis of cost and on the grounds that the thermal storage system can act as a "buffer" to protect the turbine from rapid variations in the solar energy input. The electrical energy generated by photovoltaic cells would most likely be stored in batteries, but could also be converted and stored in the form of mechanical or chemical energy. The mechanical energy intercepted by wind turbines may be stored in spinning flywheels, by pumping water, or by compressing air, or it may first be converted into electricity and then stored by other means.

The storage of energy for use in the transportation sector will also play a crucial role in the future. Among the foreseeable options, vehicles powered by a combination of electrical batteries, hydrogen fuel, and fuel alcohols derived from biomass appear to offer the most favorable approach.

The long-term need for storage may be reduced to some extent by changing consumption patterns in order to make them more closely coincide with the availability of energy (Metz 1978). This could be accomplished by a number of methods, generally with minimal inconvenience. For example, computer-operated load management could allow that certain tasks—relatively insensitive to timing—be satisfied when energy supplies are abundant. Load shifts could also be encouraged by measures such as time-of-day pricing.

Of even greater significance, a judicious choice of energy-generating facilities could further reduce the degree of storage capacity required. For example, in areas where a good correlation exists between building heat loads and high wind velocities, a system relying on wind energy to provide space heat (optimally via heat pumps) would require limited storage (Heronemus 1976).

Widely distributed solar- and wind-electric plants feeding power into an integrated utility grid would tend to cancel out power fluctuations due to local variations and thereby reduce overall storage requirements. One study by Justus (1976) concluded that a large array of interconnected wind turbines could supply electric power with 95 percent reliability if twenty-four to forty-eight hours of storage capacity were built into the system. However, a thorough assessment of the economic and environmental trade-offs between the installation of additional storage or energy transmission capacity should be a major consideration in the design of future energy systems. Sorensen (1976) concluded that Danish wind generators with ten hours storage are as reliable as nuclear power plants.

Coupled solar- and wind-electric systems also appear attractive. On a seasonal basis in the Northern hemisphere, solar insolation is more intense in summer than in winter, whereas average wind velocities tend to be greater during the winter (Putnam 1948). A computer model developed by Andrews (1976) suggests that energy storage capacity might be reduced by a favor of two to four in a coupled solar-wind system.[2]

Regardless of the energy production facilities ultimately selected, energy storage is destined to play an increasingly important role in the future. A multitude of options are currently under development, and undoubtedly, novel concepts will be devised with time. A more detailed technical discussion of the major storage systems is presented in the following section.[3]

2. This result was obtained assuming a single location. The geographical dispersion of numerous interconnected solar and wind arrays, which would further reduce storage requirements, was not considered in this model.

3. Readers not interested in the specific details might consider skipping over to the "Overall Assessment" section.

Technology

Storage technologies may be subdivided into thermal, thermochemical, electrochemical, mechanical, and electrical systems, depending on the manner in which energy is stored.

Thermal Storage.   Thermal storage is the most appropriate storage
mode whenever the energy source or end use form is thermal energy.
Thermal storage is needed for both low and high temperature applications. Thus, the storage technique must be capable of operating in
the temperature range of the energy collected and that needed at the
point of end use. For space conditioning and water heating, relatively
low temperature stores are the most practical. For thermal power
conversion systems, high temperature storage is needed in order to
obtain reasonable thermodynamic conversion efficiencies.

Low temperature thermal storage systems are quite flexible and
could be incorporated into heating systems in a variety of ways.
In the simplest case, a storage system could be charged when excess
energy is available and discharged when thermal energy is needed.
For example, thermal energy collected by solar collectors during the
day could be temporarily stored and used at night. As another example, off-peak electric power could be used to operate an electrically
driven heat pump to charge intermediate temperature storage systems. The thermal energy could then be used during peak load
periods. This could alleviate utility "load-leveling" problems and
could thereby contribute to reduced electricity costs.

The development of reliable and economical seasonal storage
systems could lead to the more rapid installation of solar energy
systems capable of supplying total annual heating requirements. In
these systems, excess solar energy would be collected during the
summer and used for heating during the winter. A scale comparable
to community size heating plants or larger now appears most attractive for such systems. Large-scale, seasonal storage facilities offer
significant economic advantages over systems designed for individual
dwellings. If water reservoirs are used for seasonal storage, the
volumes required will be rather large. One approach that has been
suggested for multiunit storage systems involves the storage of hot
water in underground caverns or aquifers (OTA 1978). Underground
storage using the heat capacity of the earth has also been proposed,
and several prototype systems are under development (Green et al.
1976; Margen 1978; Riaz et al. 1976). Another system capable of
providing seasonal storage is the Annual Cycle Energy System
(Fischer 1975). In this system, a low temperature water tank is used
as the thermal source for a heat pump during the winter. The water

freezes as energy is removed from the tank. The ice is stored and used in the summer for air-conditioning purposes. This concept is reminiscent of ice houses, which were widely used in this country before the introduction of electrical refrigeration.

Thermal storage can be accomplished in either or both of two ways: sensible heat storage—where thermal energy is stored by raising the temperature of the storage medium—or latent heat storage—where the energy is stored by a phase change in a material which is a physical change of state such as simple melting and freezing either requiring the addition of heat or involving the release of heat. Sensible heat storage has the advantage of dependability and, in general, an inexpensive storage medium. On the other hand, the material employed by this storage method is likely to be bulky.

The most suitable phase changes for thermal energy storage involve solid-liquid transformations such as the melting of a pure substance or the melting of salt hydrates in their water of crystalization. Because the heat of fusion of most materials is much larger than the amount of sensible heat involved in practical temperature swings, latent storage has the advantage of compactness and may be especially useful in applications where the volume and/or weight required for a sensible heat storage system may be prohibitive. One disadvantage of latent storage is that the heat exchanger must be integrated into the storage device, thereby complicating system design. A suitable phase change material should be noncorrosive, chemically stable, and nontoxic and have a large heat of fusion and a melting point within the needed suitable temperature range. It should also be capable of sustaining prolonged cycling.

The primary candidate materials for low temperature sensible heat storage are water and rocks. Normally, water storage is used in conjunction with water-heating systems and rocks with air systems. The main advantages of water storage are the high specific heat (1.0 Btu/lb°F) and the low cost of water. Rocks have a lower specific heat (roughly 0.16–0.20 Btu/lb°F). Consequently greater mass (by a factor of about five to six) and volumes (a factor of two to three) are needed to store the same amount of thermal energy in rocks as in water. However, rocks are more easily contained than is water, and systems employing rocks and air circulation avoid the problems of freezing and corrosion associated with liquid systems. Rocks can also act as heat exchangers, thereby reducing overall system costs.

Water may be stored in tanks at a cost of about $1 to $4 per thermal kilowatt hour (kWht) (Oakes 1977; Pickering 1975), assuming a temperature swing between 50° to 100°F. Hot water storage in underground caverns and aquifers can be much cheaper, with costs estimated as low as $0.03/kWht (OTA 1978). One large-scale experiment with aquifer storage systems is currently being conducted near Bucks, Alabama. The cost of rock bed storage has been estimated at about $0.20–1.00/kWht (Kreider and Kreith 1975).

High temperature sensible heat can be stored in steam, oils, molten salts, liquid and solid metals, and a host of other solid materials. Costs estimated for a number of these storage systems range from about $3.30–8.00/kWht (EPRI 1976; Green et al. 1976; Kalhammer and Schneider 1976). Intermediate to high temperature underground storage using earth materials has also been proposed. Long-term underground storage at a temperature of 930°F has been estimated to cost anywhere from about $0.03/kWht (OTA 1978) to $0.20/kWht (Riaz et al. 1976). In light of the factor of 100 cost differential between these systems, underground thermal storage clearly offers the preferable approach if cost predictions are born out by experience.

The lowest cost thus far estimated for thermal storage has been about 3 cents per kWht of storage capacity. As a point of comparison, conventional energy forms such as oil and gas typically cost somewhat less than 1 cent per kWht. This analogy is deceptive, however. As explained in the final section of this chapter, the overall economics governing energy storage involves a subtle equation incorporating a broad range of parameters. The cost cited above for the storage system represents a "fixed" cost that is amortized over a period of time. After this cost is defrayed, storage is essentially free. Therefore, the operating cost per unit of energy stored turns out to be considerably less than the fixed capital cost of installing storage capacity. As shown later in the chapter, operating storage costs for some proposed storage systems may turn out to be only a fraction of the unit energy costs of conventional fuels and electricity.

The materials most actively considered for low temperature latent heat storage are salt hydrates and paraffins. Among the salt hydrates, sodium sulfate decahydrate has received the most attention. It is noted for its low cost. Energy is absorbed when the solid material liquifies at its melting point of 90°F. Thermal energy may be retrieved by lowering the temperature of the storage medium below the melting point, so that all or part of the material reverts to solid form. The major problem encountered with the use of this material, and some other salts as well, is one of unreliable melting and recrystallization. Efforts have been made to overcome this problem using thickening agents, stirrers, and nucleating agents (Telkes 1975) and more recently by the encapsulation of the storage medium in rolling cylindrical drums (SEIR 1977). This problem can be avoided by using certain other salts, such as potassium fluoride tetrahydrate

(Schroder 1974), or by using a single, pure material such as paraffin. However, paraffins and other organic compounds are relatively expensive. Cost estimates for a number of storage systems using salt hydrates and paraffins range from about $3-8/kWht and $13-32/kWht, respectively (Joy and Shelpuk 1976; Telkes 1975).

Thermal energy can also be stored in phase-changing eutectic salts and metal alloys at temperatures exceeding 750° F. Cost estimates for systems employing a suitable electric salt and alloy are $3.14/kWht and $4.55/kWht, respectively (Kaufman and Lorsch 1976).

Thermochemical Storage. Energy can be stored and recovered by use of reversible chemical reactions, as biomass or a biomass-derived fuel such as alcohol and methane, or in the form of hydrogen. This section will focus on the first approach, as the other methods—biomass and hydrogen—are treated separately (in Chapter 4 and in the final section of this chapter, respectively).

The technology for energy storage using chemical reactions is just beginning to be explored. The basic principle is as follows: thermal energy is applied to a chemical capable of undergoing a reversible, endothermic reaction. The reaction products would be separated and stored. They could be transported to a distant recombination center, if required. When thermal energy is needed, the dissociation products could be combined to produce the reverse, exothermic (energy-releasing) reaction. In order to be suitable, the reactions must occur at practical temperatures and be reversible.

The storage of thermal energy employing endothermic (energy-absorbing) chemical reactions provides features not available with sensible and latent heat storage. Because heats of reactions can be considerably larger than sensible heats or heats of fusion—up to ten times greater per unit volume according to Simmons (1976)—chemical storage has the potential for achieving higher energy densities. Thermal energy can be chemically stored at ambient temperature for an indefinite period of time without the need for insulation and with no thermal losses. Stored energy may be recovered at temperatures above or below the input temperature. Another advantage is that the stored energy in the form of the reaction products would be easily transportable, allowing for flexibility in storage arrangements. For example, chemical heat pipes, which could be used to transport solar thermal energy temporarily stored as chemicals over distances of hundreds of miles, are currently under study. These systems appear to be economically competitive with gas pipeline transportation and are also attractive from an efficiency point of view, with overall values of 80 percent anticipated (C&EN 1979). Chemical storage may ultimately prove to be suitable for a variety of applications ranging from low temperature space heating to utility storage for thermal power plants.

Hundreds of potentially useful reactions are currently under study. A few of the more noteworthy will be briefly mentioned. The "Solchem" system is based on the decomposition of sulfur trioxide into sulfur dioxide and oxygen when thermal energy is added at temperatures of 1,470° to 1,830°F. Thermal energy can be recovered by recombining the sulfur dioxide and oxygen in the presence of a catalyst. An overall efficiency of 90 percent may be achievable for this system (Offenhartz 1976). The "ADAM-EVA" system is based on a familiar reaction involving conversion of methane and steam into carbon monoxide at a temperature of about 1,830°F (Kalhammer and Schneider 1976). A system based on the conversion of nitrogen tetraoxide into nitrogen dioxide may be attractive owing to low operating temperatures in the 23° to 212°F range (Grodzka 1975). Low material costs are expected for reactions involving magnesium or calcium oxides at temperatures of about 700°F and 970°F, respectively (Ervin 1975). Reversible reactions based on the decomposition of potassium, sodium, and barium oxides are considered promising because of potentially high energy densities (Simmons 1976). Metal hydrides, being tested for the storage of hydrogen, can also provide a means of storing thermal energy at low to intermediate temperatures. High storage densities and low material costs are predicted for hydride systems (Davison 1975; Gruen and Sheft 1975).

However, chemical storage is at present an untried concept whose economics are uncertain. The potential advantages are nonetheless sufficient to warrant a vigorous research and development program in order that the prospects for this storage technology can be better assessed.

Electrochemical Storage—Electrical Batteries.   Batteries are devices that store energy in chemical form, although electrical energy is both the input and the output form. There are two generic types, primary and secondary. Primary batteries produce electricity until their original material is consumed. The reaction is not reversible. In general, the electrodes must be replaced if these batteries are to be reused. Secondary (or storage) batteries, on the other hand, can be recharged for future use. During charging, direct current electricity is electrochemically converted to chemical energy. During the discharge cycle, most of this energy is later converted back to direct current electricity.

Storage batteries have several attractive features. They have all the advantages of modularity; virtually any amount of energy can be stored, and any voltage or current can be achieved by interconnecting the separate cells in suitable arrangements. Furthermore, batteries have no moving parts and emit no noise. No pollution is generated during operation, although some may be produced during manufacturing processes. Batteries have no specific siting requirements and consequently can be placed close to their loads, thereby reducing demands on transmission lines. Unfortunately, currently available batteries are too expensive and generally have energy densities that are too low for widespread use.

The improvement of electric batteries for vehicular propulsion is the object of an intensive developmental effort. Satisfactory batteries could play an extremely important role in powering the transportation sector in a renewable energy economy, as discussed further in Chapter 6.

Desired characteristics of storage batteries are low production costs, high energy and power densities (to reduce storage volume and mass), high durability, and long cycle life. Charge-discharge efficiencies should be about 75 percent or higher, and the chemical reaction should be highly reversible. Especially important for the performance of electric vehicles are a high specific energy and specific power. The specific energy, expressed in watt hours per pound (Wh/lb), is the dominant parameter determining the maximum range. The specific power, expressed in terms of watts per pound (W/lb), measures the rate that energy can be extracted per unit mass of the battery and governs the acceleration and maximum speed achievable by the vehicle. General targets for advanced batteries are a specific energy of 100 Wh/lb, a specific power of 100 W/lb (JPL 1975) and a cycle life of 2,000–5,000, or equivalently, ten to twenty years (EPRI 1976). Specific energies projected for all batteries are one to two orders of magnitude below that of liquid petroleum, which has a value of about 6 kWh/lb or, equivalently, 6,000 Wh/lb. Currently available storage batteries cost more than $60/kWhe. Projected costs for many advanced batteries fall within the range of $10–40/kWhe of capacity, which, if achieved, could make them economical for a broad range of storage and transportation applications (EPRI 1976; Kalhammer and Schneider 1976; OTA 1978).

Resource scarcity will probably make material recycling imperative and force a reliance on several different batteries, suitable for different applications, rather than dependence on a single system. Several of the most important options are briefly reviewed here.

Lead-acid batteries are the most familiar of the storage batteries, widely used today in automobiles for starting, lighting, and ignition.

These batteries suffer from an inherently low specific energy and a high cost. The specific energy of current batteries is about 14 Wh/lb. Modifications such as replacing the solid lead electrodes might improve this value to 23 Wh/lb by 1980 (C&EN 1977), still relatively low in comparison with other advanced batteries. At present, lead-acid batteries cost about $65/kWhe, although advanced cells may be provided in the future for $25/kWhe (Birk and Kalhammer 1976). A major constraint to the extensive use of this battery is posed by the limited availability of lead.

Nickel battery systems share some of the drawbacks of lead-acid batteries. They suffer from a relatively low specific energy and from resource constraints that should limit their utilization to an interim period until the introduction of more advanced batteries. Within this class, both the nickel-zinc and nickel-hydrogen batteries are considered to be strong candidates for storage and motive applications (Gross 1976).

The zinc-chlorine battery appears promising for both utility storage and automotive applications and should become commercially available within a few years (C&EN 1977; Gross 1976). Good range and performance have been reported in a prototype test vehicle operated on a zinc-chlorine battery. It has been estimated that these batteries could be mass produced at costs between $25/kWhe (EPRI 1976) and $40/kWhe (OTA 1978).

The zinc-bromide battery, a relative of the zinc-chlorine cell, is also being developed for utility load leveling and electric vehicles. Experimental devices of this type have been operated for more than 2,000 cycles. Efficiencies of 65 to 80 percent and costs of $17-26/kWhe have been predicted (OTA 1978).

Redox battery systems hold great promise for large-scale electricity storage owing to their potentially low costs. Preliminary estimates range from $5-35/kWh (EPRI 1976; Kalhammer and Schneider 1976; OTA 1978). Iron-redox batteries may be producible for as low as $12/kWhe (OTA 1978).

High temperature batteries are of interest because they offer the promise of long lifetimes and high energy densities at potentially low costs. However, the elevated operating temperatures pose safety and material problems and also necessitate highly effective thermal insulation. These factors pose practical problems for the use of high temperature batteries in electric vehicles and may make electric utility storage a more attractive application, although both applications are currently under study. The sodium-sulfur battery is the most advanced in this group. Future costs are estimated at $15-30/kWhe (Kalhammer and Schneider 1976). A production cost of $24/kWhe has been predicted by 1988 (OTA 1978). A cost of $15–25/kWhe has been projected for the sodium-chloride battery, which operates at a slightly lower temperature (EPRI 1976). Lithium-metal sulfide (sometimes called lithium-sulfur) batteries are expected to be available for commercial use by the mid-1980s at a cost of $20–30/kWhe (Kalhammer and Schneider 1976; OTA 1978; SRI 1976).

Air batteries use oxygen from the air as the positive electrode (cathode). Because the oxygen does not have to be stored, a potentially low battery weight and correspondingly high specific energy is possible. Both aluminum-air and iron-air batteries look attractive for use in electric vehicles (Gross 1976). The most promising battery of this class, however, is the lithium-water-air battery. This is a primary battery. Preliminary research indicates a potential for a high specific energy (200 Wh/lb) and a high specific power (91 W/lb) (O'Connell et al. 1975). The most attractive strategy for implementation appears to be a "hybrid" system in which the primary battery is complemented with a secondary storage battery.[4] It has been estimated that a one ton automobile powered by a 500 pound primary-secondary battery system could achieve a sustained speed of 60 mph and a range of up to 1,000 miles (O'Connell et al. 1975). "Recharging" would take place by the periodic replacement of the lithium anode. However, the availability of lithium may pose a problem, especially in the event of competition from a nuclear fusion program employing lithium to "breed" tritium. Therefore, the recycling of lithium from replaced anodes would be mandatory.

Mechanical Energy Storage.

Pumped Hydroelectric Storage. Pumped hydroelectric storage is a well-established technology and is the only storage method currently in use for large-scale utility applications. This system consists of two reservoirs separated in elevation by a "head," generally of 300 feet or more. During off-peak hours, excess baseload electric capacity is used to pump water from the lower to the upper reservoir, thereby storing gravitational potential energy. During periods of peak demand, the elevated water is allowed to flow to the lower level through a turbine generator, where the kinetic energy of the flowing water is converted into electricity. Pumping and generation may be accomplished by using separate pumps and turbines, or the two operations may be combined using a reversible pump-turbine connected to a

4. Battery-flywheel hybrid vehicles are discussed later in this chapter.

motor-generator. Efficiencies for the overall process normally fall within the range of 66 percent (FPC 1971) to 75 percent (Allen 1976).

Windmills have been used to pump water since ancient times and could readily be linked to a pumped storage facility. The mechanical energy intercepted by wind turbines may be used to pump water directly or may be converted to electricity, which could then be used to operate pumps; the second option would have a somewhat lower overall efficiency. Large pumped storage reservoirs could also then be used as buffer systems to smooth out power fluctuations from solar and wind-electric generators. In this mode of operation, the output of the hydroelectric power station would be regulated to account for fluctuations in solar and wind generation and the demand for energy. Installed hydroelectric generating capacity in the United States presently stands at 59,000 MWe, with an additional 10,000 MWe of pumped storage capability. It has been estimated that at least 10 percent, or a total of 6,000 MWe, of the currently installed power generation capacity could be utilized for storage purposes (Metz 1978). Further major expansions, however, would likely be constrained by environmental, siting, and safety problems.[5]

Major environmental problems would result from creation of the reservoirs. Impacts would also result from the operation of the plants, because water levels in the reservoirs would be constantly changing, and water from different portions would be continually mixed by pumping (OTA 1978).

Pumped hydro plants are only considered feasible on a large scale; the minimum size considered economical is about 10,000 MWhe (APL 1975) or roughly 200 to 2,000 MWe (EPRI 1976). The large plant size contributes to long lead times and large capital requirements for construction. In general, these facilities would have to be located far from load centers, necessitating long distance transmission of electricity.

The limited availability of suitable sites and the large land and water requirements will inhibit the growth of this technology. The use of underground water storage reservoirs and generators could alleviate, to some extent, the topographical restrictions placed on conventional plants. Natural underground caverns or abandoned mines have been suggested for housing the underground water. Although a large number of suitable sites are available for underground facilities, the need for satisfactory rocks at the required depths and for conditions appropriate for the surface reservoir may in fact limit

5. Also discussed in Chapter 4.

these sites to a smaller geographical area than for conventional plants (Allen 1976).

One attribute of pumped storage is a long useful plant lifetime. Factors affecting plant life are possible silting of the reservoirs and degradation of the dam and generating equipment. The expected lifetime for these systems ranges from fifty years (Allen 1976) to one hundred years (Corso 1976).

The largest pumped storage plant operating in the United States is located in Ludington, Michigan. The planning and construction of this plant took fourteen and one-half years to complete at a total cost of about $340 million (Robinson 1974a). The storage capacity of the Ludington facility is about 15,000 Mwh and the maximum rated power is 1,900 MWe (Robinson 1974a). The unit storage costs are thus about $23/kWhe or $179/kWe. The estimated capital costs of future pumped hydro plants is $240–320/kWe, depending on the site and size of the plant (Allen 1976), and $270–350/kWe for underground facilities with ten hours of storage capacity (Scott 1977).

Compressed Air Storage. In compressed air systems, excess power is used to compress large quantities of air during periods of low demand and to pump it to a storage reservoir. During peak demand hours the compressed air is used to run a turbine, either with or without the burning of a fuel. Although compressed air storage was first proposed in the 1930s, it has only recently attracted serious attention. Compressed air storage is a candidate for large-scale utility storage and offers greater siting flexibility and imposes smaller land area requirements than pumped hydroelectric storage.

The air storage reservoir must be airtight and capable of withstanding pressures on the order of 40 atmospheres (600 psi). Natural reservoirs such as caverns, aquifers, and depleted oil or gas wells may be suitable. Caverns can also be artificially formed by the process of solution mining wherein the storage reservoir is created by injecting fresh water into a salt formation. Salt is leached, and the saturated brine is removed, creating a cavern. Of all the different approaches, the storage of compressed air in hard rock caverns appears to have the greatest potential, due to the existence of suitable hard rock at the required depths throughout most of the United States (Hobson et al. 1976). Other manmade reservoirs, such as steel tanks or abandoned mines, could also be used, but are generally expected to be too costly.

The application that is currently attracting the greatest developmental effort is the combination of compressed air storage systems with gas turbines. This approach may become practical in the near future. "Conventional" hybrid plants will be large and are only expected to be economic on the scale of 200–2,000 MWe (EPRI 1976). In such an arrangement, electricity would be used to compress air to a high pressure. The sensible heat of compression would be removed by means of intercoolers, so that the compressed air could be stored at near ambient (geothermal) temperatures, to alleviate thermal stress on the storage reservoir. The loss of thermal energy reduces the overall efficiency of this system. During periods of peak demand, air would be withdrawn from storage, heated by the combustion of a fuel, and then expanded through a gas turbine generator. In a normal gas turbine cycle, roughly two-thirds of the power output from the turbine is required to drive the air compressor. The use of already compressed air would, in theory, triple the net power output. In practice, however, the output would only be increased by a factor slightly greater than two (Bush et al. 1975; Kreid 1976). The overall storage efficiency is defined here as the ratio of the efficiency of the compressed air-gas turbine system compared to that of the baseload generating plant. It may range from 47–74 percent (Kreid 1976; EPRI 1976) or lower, depending on the amount of heat energy lost (Glendenning 1975).

The combined compressed air-gas turbine system is considered to be only a few years away from introduction in the United States (Chang 1977). One hybrid system has been built in Huntdorf, West Germany (Stys 1975). The capital cost of this plant is about $40 million, and its power rating is 290 MWe (Ricci 1975), translating to a unit power cost of $138/kWe. Advanced systems, utilizing heat recuperation, modern turbine design, and perhaps fluidized bed combustion, are under study and may be available in the 1980s (Chang 1977). Cost projections for future combined cycle plants range from $120–176/kWe (Hobson et al. 1976).

The hybrid concept is not a pure energy storage system. Compressor energy is deliberately dissipated as thermal energy, and in addition, the operation of this system is dependent on the use of a fuel such as oil whose supply is limited. Consequently, its usefulness may at best be limited to an interim period.

The major shortcomings of conventional, fuel-fired compressed air storage could largely be avoided in adiabatic systems in which the thermal energy generated by compression is stored. In this system hot compressed air would be passed through a crushed rock storage bed, transferring its thermal energy to the bed. The cooled air would then be pumped to the air storage reservoir. The air would later be heated by this stored thermal energy before passing through the turbine. Estimates for overall storage efficiencies (i.e., percentage of input electricity recovered in the turbine generator) for this system varies from 50 percent (Kreid 1977) to 75 percent (Stephens 1975; Glendenning 1975).

The adiabatic storage approach has only recently been looked into. Suitable equipment is not available today, so that new machinery would need to be developed before these systems could be deployed. Introduction is expected to be at least five years off (Chang 1977). Owing to the need for thermal storage and the larger-sized components resulting from the lower energy and power densities of the non-fuel-burning systems, the capital costs for the adiabatic systems are certain to be higher than those for conventional gas turbine hybrid systems. One preliminary cost estimate for a no fuel system using a solution-mined reservoir was $220/kWe (Glendenning 1975). However, operating costs would be lower than those for conventional systems, so that adiabatic compressed air storage is likely to be competitive with fuel-fired systems as soon as they are ready for introduction. Adiabatic systems may prove to be an attractive large-scale storage option and are worthy of a vigorous research and developmental effort.

Flywheel Storage. The concept of storing energy in flywheels is not new and was in fact utilized in potter's wheels several thousand years ago. Flywheels store kinetic energy in a rotating mass, the amount of energy being proportional to the moment of inertia of the wheel and the square of the angular velocity of the flywheel. The principal is simple: in a power-generating system, the flywheel is coupled to a spinning shaft with a clutch. During periods of excess power, the flywheel is spun up to high velocities. The stored energy is later drawn off to drive electric generators or used to perform useful work. The flywheels would have to be enclosed and run in a vacuum to reduce the amount of energy lost to viscous heating of the air.

The major applications proposed for flywheels are electrical storage and vehicular propulsion. Flywheels may be used to recover the energy normally dissipated as heat during braking. Flywheel braking systems have been employed on experimental subway trains in New York City and on electric trolleys in San Francisco. One study found that flywheel braking systems could improve the fuel economy of automobiles by approximately 10 percent (Gelb et al. 1972). Flywheels are expected to have a high specific power—greater than 100 W/lb (Robinson 1974b)—which is important for acceleration in electric vehicles. They can be recharged in a period of a few minutes, while batteries may require several hours. However, flywheels will probably not be able to match the specific energy of the best advanced batteries. Therefore, a flywheel-battery hybrid system might be an attractive option. One study found that the use of flywheels in conjunction with electric batteries can considerably boost system power capability while keeping within vehicle weight constraints (Behrin, et al. 1977). An ERDA (1975) report indicated that the use of flywheels in automobiles, in conjunction with a heat engine or an electric battery, could lead to savings of oil on the order of 100 million barrels per year by 1995. The same report found that flywheel storage in the utility sector could reduce the cost of generating electricity and might save 220 million barrels of oil (or an equivalent fuel) over the 1985-1995 period.

Flywheels are quite flexible in scale and can be sized according to the application. The storage capacity of flywheel systems may be only a few kWhe or as large as 100 MWhe or more (ERDA 1975). The nonpolluting nature of flywheels, combined with their load-matching capability, makes possible their location near the point of end use.

Until now, most flywheels have been composed of steel. The operational hazards of flywheels are proportional to their specific energy, so that the usual practice has been to operate them at relatively low speeds rather than attempting to provide for adequate containment in the event of a failed flywheel. As a result, while the theoretical maximum specific energy of a modern steel flywheel is about 26 Wh/lb, safety considerations limit ratings to normally about 12 Wh/lb and less than 1 Wh/lb for very large flywheels (APL 1975).

The new approach to flywheel design is based on the use of unidirectional (anisotropic) materials generally composed of fiberglass, epoxies, carbon fibers, wood, and organic materials (such as kevlar). These materials are roughly ten to twenty times stronger than steel (Blackstone 1975). Consequently, flywheels made of these materials can be spun at much higher speeds, storing greater amounts of energy. Tests suggest that these new "superflywheels" would be much safer than those made of steel (APL 1975). Advanced flywheels are expected to achieve specific energy values of 60-70 Wh/lb by 1985 (ERDA 1975) and charge-discharge efficiencies of 70-85 percent (EPRI 1976). Future costs of $69/kWe for one hour storage and $285/kWe for ten hours of storage capacity have been projected (APL 1975).

Electrical Storage—Superconducting Magnets. Storage using superconducting magnets is the only technique being pursued today for directly storing electrical energy without conversion to other forms.

Photograph 5-1: Flywheel Energy Storage. Artist's conception of an energy storage facility with six flywheel-driven 2.5 megawatt generators. The prime mover of each generator, located below ground level, consists of four tandem 95-inch radius wheels weighing approximately 62 tons each. Each generator would be capable of storing 19,000 kilowatt-hours of electrical energy.

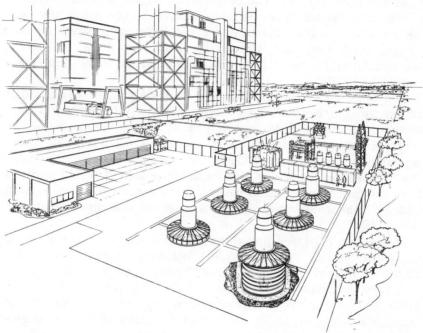

Source: Rockwell International.

The energy is stored in a magnetic field by circulating a current through the windings of an electromagnet. In a conventional electromagnet, electrical resistance results in power losses so that power must be continuously supplied to maintain the magnetic field. However, if the coil winding is composed of superconducting materials (such as niobium, titanium, and lead) and is cooled to extremely low temperatures, near absolute zero, it becomes "superconducting" as the resistance drops to zero. Therefore, once the magnetic field is established, no additional power is needed to maintain it. The energy could be stored in a resistanceless inductor for an indefinite period. In normal operation, the magnet is charged during periods of excess available power. The stored energy is recovered as needed.

Because the electricity is not converted to other forms, the percentage of electricity recovered from storage would be quite high, about 95 percent (Robinson 1974b). However, alternating current electricity would first have to be converted into direct current before being fed into the magnet and then reconverted back to alternating current after withdrawal from storage. Energy would also be required for the refrigeration needed to maintain the low operating temperatures, and additional losses would be incurred in the voltage transformer. It is estimated that 10-20 percent of the input energy would be lost due to power conditioning, refrigeration, and voltage transformation (EPRI 1976). Therefore, overall efficiencies are estimated to range from 80 to 90 percent (Kalhammer and Schneider 1976).

The intense heat generated in the event of a failure of the superconductor (i.e., circumstances resulting in the system becoming nonsuperconducting) could pose a severe safety hazard. It appears that this problem can be circumvented by using "cryogenically stabilized" superconductors (Robinson 1974b). Cryogenic stabilization means that if the superconductor were to fail, the current would be drawn from the cables by the metal backing formed of copper or aluminum, and the heat would be dissipated throughout this support structure (Hein 1974).

It is generally agreed that this storage method would only be economically feasible on a large scale. Storage capacities on the order of 10,000 MWh have been suggested (Kalhammer and Schneider 1976).

A very preliminary cost estimate of $500-600/kWe has been made (FEA 1974). Additional developmental work will be needed to determine a more precise figure. Although considerable technological advances are required before commercialization can be considered, the potentially high efficiencies of superconducting storage make it worthy of consideration as a contender for large-scale electrical utility applications.

Overall Assessment

In the future, a mix of storage technologies will almost certainly be required for a broad range of applications. The current state of the art and future prospects for the various systems are summarized here. Estimates of the state of commercial availability, efficiency, costs, and other important physical and economic characteristics are presented in Table 5-1.

Energy storage costs quoted in the general literature are normally presented in terms of dollars per kilowatt or dollars per kilowatt hour of storage capacity. These simple formulations neglect characteristics—such as the anticipated lifetime, efficiency, and capacity factor of the storage systems—that have a direct bearing on the overall economics of various energy storage technologies. Therefore, in order to facilitate meaningful comparisons between different systems, storage costs per kilowatt hour of energy stored over the entire lifetime of each system have been computed, based on a range of reasonable and mutually consistent assumed values for the relevant parameters.[6] Results are presented in the "Energy Storage Cost" column of Table 5-1. Although the actual cost computations are subject to major uncertainty, they nonetheless can provide rough indications of the relative attractiveness of the competing storage systems.

From Table 5-1, one sees that the estimated costs of energy storage by some methods would amount to only a small fraction of the current costs of conventional energy forms, whereas storage costs for other approaches may constitute a significant fraction or even exceed present conventional energy costs. For example, estimated thermal storage costs range from 0.03 to 8 mills per kWhe. In comparison, the 1979 costs of residential heating oil and natural gas

6. Costs have been computed by the standard formula used in crude calculations of busbar electrical generation costs:

$$\text{Energy Storage Costs (mills/kWh)} = \frac{(\$/\text{kW}) \; (FCR)}{(\text{Efficiency}) \; (8.76 \; \text{hr.}) \; (\text{Capacity Factor})}.$$

The fixed charge rate, *FCR*, was determined by using the equation:

$$FCR = \frac{(1 + r)^T \; (r)}{(1 + r)^T - 1},$$

where r represents the interest rate (assumed to be 0.15) and T represents the expected useful life of the system (in years). These assumptions yield typical values for *FCR* generally ranging from 0.15 to 0.20. This is admittedly a simplistic method of computing *FCR*, which, for utilities, is in practice a complicated variable reflecting several other considerations, including the cost of borrowing capital, the debt-to-equity ratio, and income taxes. Nonetheless, all systems have been treated in a mutually consistent manner.

The capacity has a profound effect on overall costs, but cannot be confidently predicted for systems that have not yet been built. Therefore, a value of 0.25, or 25 percent, has been assumed for all storage systems. This corresponds to the average load factor of presently operating pumped storage plants (FPC 1971).

Operation and maintenance costs have not been included in these computations. These costs cannot be accurately predicted for systems other than pumped hydroelectric storage, but are expected to be relatively insignificant. Operation and maintenance costs for pumped storage facilities have averaged only about 0.1 to 0.3 mills per kWh (EPRI 1976).

Pumped hydro storage is treated somewhat differently, as explained in Table 6-1.

Table 5-1. Storage Technologies.

Technology	Date of Introduction	Specific Energy (Wh/lb.)	Volumetric Energy Density (kWh/ft³)	Lifetime (years)
Thermal				
Low temperature Heat				
Sensible	present[c]	3-29[l]	0.03-1.8[v]	20-30[dd]
Phase change	1990[d]			20[ee]
Hydrates		16-35[m]	1.8-3.4[w]	
Paraffins		18-26[n]	0.9-1.3[w]	
High Temperature Heat				
Sensible	before 1885[e]	26-79[o]	3.1-4.6[x]	20-30[e]
Phase change	1990[d]	21-110[p]	7.6[y]	20[w]
Chemical (Hydrogen)	1985-2000[f]	117-586[q]	10-27[z]	20-25[e]
Electrochemical				
Lead-Acid Batteries	present[g]	14-23[e]		5-10[e]
Advanced Batteries	1985-2000[d,e]	25-400[s]		10-20[e]
Mechanical				
Pumped Hydro	present[h]	0.12[t]	0.008-0.04[aa]	50-100[ff]
Compressed Air				
Gas turbine	present[i]		0.1-0.5[bb]	20-25[e]
Adiabatic	1985[j]		0.5[cc]	20-25[e]
Flywheel	1985-2000[k]	20-70[u]	0.5-3.0[d]	20-30[gg]
Electrical				
Magnetic (SMES)	after 2000[e]			20-30[e]

[a]"(e)" and "(t)" refer to electrical and thermal; storage costs computed in terms of electrical kilowatt hours have also been placed on a thermal primary energy equivalent basis.

[b]One mill equals 1/10th of one cent, 1/1000th of one dollar.

[c]Low temperature sensible heat storage in water and rocks is widely used today in residential solar heating and conventional hot water systems.

[d]McCaull (1976).

[e]EPRI (1976).

[f]Dates given specifically for hydrogen energy storage systems (EPRI 1976). Other thermochemical storage systems may also be introduced within that period.

[g]Lead-acid batteries are used in the majority of automobile engines today for starting, lighting, and ignition.

[h]Pumped hydroelectric storage is the only large-scale storage system in use today. The first underground pumped hydro plant is expected to be installed by 1992 (Kalhammer and Schneider 1976).

[i]Construction was completed on a compressed air storage plant in Huntdorf, West Germany, in 1977.

[j]Chang (1977).

[k]Some flywheel systems are presently available. Advanced systems are expected to be ready after 1985 (EPRI 1976; McCaull 1976).

[l]This range applies to water and rock (assuming a specific heat of 1.0 and 0.20 Btu/lb°F) storage for a 50-100°F temperature difference.

[m]Assumes a heat of fusion for inorganic hydrates ranging from 56-121 Btu/lb (Lorsch et al. 1975).

[n]Assumes a heat of fusion for various paraffins ranging from about 63-89 Btu/lb (Lorsch et al. 1975).

[o]Storage capacity for a variety of sensible heat storage materials including water, oil, liquid metals, rocks, iron, and metal alloys for temperature swings ranging from 200-900°F (Green et al. 1976).

[p]Assumes heats of fusion for selected phase-changing salts and alloys ranging from 72-377 Btu/lb (Kauffman and Lorsch 1976).

[q]Davison (1975); the upper value is for a magnesium hydride system.

Overall Storage Efficiency (percent)	Storage[a] Cost ($/kWh)	Discharge Capacity (hours)	Capital[a] Cost ($/kW)	Capacity Factor	Energy Storage Cost[a] (mills[b]/kWh)
70-80[e]	0.03-4.0(t)[oo]	10[uu]	0.30-40(t)[yy]	0.25[fff]	0.03.-4(t)[ggg]
80[e]	3.0-8.0(t)[pp]	10[uu]	30-80(t)[yy]	0.25	3-7(t)[hhh]
70-80[e]	3.3-8.0(t)[qq]	10[uu]	33-80(t)[yy]	0.25	3-8(t)[iii]
80[e]	3.1-4.6(t)[rr]	12[vv]	37.7-54.7(t)[yy]	0.25	3-5(t)[jjj]
45-66[hh]			300-900(e)[zz]	0.25	32-146(e); 11-49(t)[kkk]
60-83[ii]	25-65[ii]	10[ww]	250-650(e)[aaa]	0.25	27-149(e); 9-50(t)[lll]
70-80[e]	10-40[ss]	10[ww]	100-400(e)[aaa]	0.25	9-52(e); 3-17(t)[mmm]
66-75[jj]			240-350(e)[bbb]	0.25	7-22(e); 2-7(t)[nnn]
47-74[kk]			120-176(e)[ccc]	0.25	32-47(e); 11-16(t)[ooo]
50-75[ll]			220(e)[ddd]	0.25	21-32(e); 7-11(t)[ppp]
70-80[mm]	69; 285[tt]	1; 10[xx]	69-285(e)[tt]	0.25	6-30(e); 2-10(t)[qqq]
80-90[e]			500-600(e)[eee]	0.25	39-55(e); 13-18(t)[rrr]

[r]The specific energy of lead-acid batteries is currently about 14 Wh/lb. With certain modifications, this value may be improved to 23 Wh/lb (C&EN 1977).

[s]The lower end of the range pertains to present nickel battery systems (SRI 1976). A figure of 150 Wh/lb is an anticipated value for advanced lithium-sulfur batteries (Gross 1976). Primary batteries can achieve even higher specific energies. A value of 200 Wh/lb is predicted for the lithium-water-air battery (O'Connell et al. 1975) and values up to 400 Wh/lb might possibly be achieved for an aluminum-air primary battery (Gross 1976).

[t]Assume a 328 foot head (Offenhartz 1976).

[u]Range estimated for new, high strength flywheels (APL 1975; ERDA 1975).

[v]For water and rock stores with a temperature difference of 50-100°F.

[w]Lorsch et al. (1975).

[x]Range for steam, oil, and liquid metal high temperature sensible heat storage (EPRI 1976).

[y]Value for molten salt latent heat storage (EPRI 1976).

[z]The lower value is for calcium hydroxide reactions; the higher value is for metal oxide reactions (Simmons 1976).

[aa]The lower estimate, from Offenhartz (1976), is for a 328 foot head. The upper estimate, from Kalhammer and Schneider (1976), is for a 3,000 foot head.

[bb]Kalhammer and Schneider (1976).

[cc]Stephens (1975): based on the total excavated volume—thermal plus air storage—per kWhe output.

[dd]The lower value is specifically for water storage (Pickering 1975). The higher figure pertains to several sensible heat storage systems (EPRI 1976).

[ee]A lifetime of twenty years was suggested as a reasonable expectation for storage systems using phase-changing materials (Lorsch et al. 1975).

[ff]Lifetime estimates for pumped hydro plants range from fifty (Allen 1976) to one hundred years (Corso 1976).

[gg]Twenty years is a lower bound estimate from EPRI (1976). The upper estimate is from APL (1975).

[hh]The low estimate assumes 85 percent efficient electrolysis, 90 percent efficient hydrogen storage, and 60 percent efficient reconversion in fuel cells; the

225

high estimate assumes 100 percent efficient electrolysis, 95 percent storage efficiencies, and 70 percent efficient fuel cells.

[ii]Birk and Kalhammer (1976).

[jj]The lower value is from FPC (1971); the upper value is from Allen (1976).

[kk]The storage efficiency is represented by the ratio of the efficiency of the compressed air–gas turbine system to that of the baseload plant. The following assumptions have been made: (1) the efficiency of the baseload plant is 38 percent (EPRI 1976); (2) the overall electrical conversion efficiency of the combined compressed air–gas turbine system (i.e., the electrical energy output divided by the sum of the fuel energy added to the baseload plant and the gas turbine) is 18–28 percent (Kreid 1976).

[ll]The lower end of the range is from Chang (1977) and Kreid (1977). The upper end of the range has been taken from Stephens (1975), Glendenning (1975), and Kreid (1976).

[mm]ERDA (1975).

[oo]The lower estimate is for aquifer storage (OTA 1978); the upper estimate is for water tank storage (Pickering 1975).

[pp]The lower limit is an estimate from Telkes (1975) for a storage system using Glauber's salt (sodium sulfate decahydrate). The upper estimate is from Joy and Shelpuk (1976).

[qq]The lower figure is for oil storage (EPRI 1976); the upper estimate is for steam storage (Kalhammer and Schneider 1976). A variety of storage materials fall within this cost band, including molten salt and rock bed stores (Green et al. 1976). High temperature underground storage using earth materials is projected to be considerably less expensive (Riaz et al. 1976).

[rr]The lower limit is for a phase-changing alloy system; the upper value refers to a eutectic salt system (Kauffman and Lorsch 1976). Twelve hours storage capacity has been assumed in both cases.

[ss]OTA (1978), EPRI (1976), Kalhammer and Schneider (1976)

[tt]APL (1975).

[uu]Ten hours discharge capacity from the thermal storage unit is considered to represent a typical duty cycle (EPRI 1976).

[vv]Refers to a specific storage system cited by Kauffman and Lorsch (1976).

[ww]Ten hours is considered to represent a typical battery discharge cycle (Birk and Kalhammer 1976).

[xx]Storage capacity specified for two distinct flywheel systems (APL 1975).

[yy]Value obtained by multiplying the storage cost (in $/kWh) by the discharge capacity (hours).

[zz]Low figure represents estimate of an electrolyzer-hydride storage/fuel cell system (Kalhammer and Schneider 1976). The higher capital cost estimate for an intermediate term hydrogen storage system is from EPRI (1976).

[aaa]Obtained by multiplying the storage cost ($/kWh) by the discharge capacity (hours).

[bbb]Allen (1976); Scott (1977).

[ccc]Hobson et al. (1976).

[ddd]Glendenning (1975).

[eee]FEA (1974).

[fff]A capacity factor of 25 percent was uniformly assumed for all systems.

[ggg]The following standard algorithm is used in all the storage cost calculations with the exception of pumped hydro storage:

$$\text{Energy Storage Cost (mills/kwh)} = \frac{CC\ (\$/kW) \times FCR\ (yr^{-1})}{8.76\ (hr/yr) \times e \times CF},$$

where CC represents the capital cost; FCR represents the fixed charge rate determined assuming the amortization of capital costs over the useful system life (T), assuming a net interest rate of 15 percent; e represents overall storage efficiency; and CF represents capacity factor.

The lower cost estimate assumes CC = $0.30/kWt, T = 30 yrs, FCR = 0.152, e = 0.80, and CF = 0.25. The higher cost estimate assumes CC = $40/kWt, T = 20 yrs, FCR = 0.160, e = 0.70, and CF = 0.25.

[hhh]The low estimate assumes CC = $30/kWt, T = 20 yrs, FCR = 0.160, e = 0.80, and CF = 0.25. The high estimate assumes CC = $80/kWt, T = 20 yrs, FCR = 0.160, e = 0.80, and CF = 0.25.

[iii]The low estimate assumes CC = $33/kWt, T = 30 yrs, FCR = 0.152, e = 0.80, and CF = 0.25. The high estimate assumes CC = $80/kWt, T = 20 yrs, FCR = 0.160, e = 0.70, and CF = 0.25.

[jjj]The low estimate assumes CC = $37.7/kWt, T = 20 yrs, FCR = 0.160, e = 0.80, and CF = 0.25. The high estimate assumes CC = $54.7/kWt, T = 20 yrs, FCR = 0.160, e = 0.80, and CF = 0.25.

[kkk]The low estimate assumes CC = $300/kWe, T = 25 yrs, FCR = 0.155, e = 0.66, and CF = 0.25. The high estimate assumes CC = $900/kWe, T = 20 yrs, FCR = 0.160, e = 0.45, and CF = 0.25.

[lll]The low estimate assumes CC = $250/kWe, T = 10 yrs, FCR = 0.20, e = 0.83, and CF = 0.25. The high estimate assumes CC = $650/kWe, T = 5 yrs, FCR = 0.30, e = 0.60, and CF = 0.25.

[mmm]The low estimate assumes CC = $100/kWe, T = 20 yrs, FCR = 0.160, e = 0.80, and CF = 0.25. The high estimate assumes CC = $400/kWe, T = 10 yrs, FCR = 0.20, e = 0.70, and CF = 0.25.

[nnn]Because the expected useful system life for a pumped hydro plant exceeds the normal amortization period, a slightly different formula is applied in which total costs accumulated over a thirty-year amortization period are divided by the net amount of energy stored over the lifetime of the system:

$$\text{Storage Costs} = \frac{(CC) \times (0.152/\text{yr}) \times (30 \text{ yr})}{(8.76 \text{ hr/yr}) \times (T \text{ yrs}) \times e \times CF} .$$

The low estimate assumes CC = $240/kWe, FCR = 0.152 T = 100 yrs, e = 0.75, and CF = 0.25. The high estimate assumes CC = $350/kWe, FCR = 0.152, T = 50 yrs, e = 0.66, and CF = 0.25.

[ooo]Low estimate assumes CC = $120/kWe, T = 25 yrs, FCR = 0.155, e = 0.74, and CF = 0.25. High estimate assumes CC = $176/kWe, T = 20 yrs, FCR = 0.160, e = 0.47, and CF = 0.25. In addition, operational fuel costs have been included. With an assumed heat rate of 5,000 Btu/kWhe for the gas turbine alone (EPRI 1976; Kreid 1976) and a fuel oil price of about $14/barrel, fuel costs are equivalent to about 12 mills/kWhe. Assuming 4 percent inflation over the twenty to twenty-five year period of operation, average fuel costs amount to about 20 mills/kWhe.

[ppp]Low estimate assumes CC = $220/kWe, T = 25 yrs, FCR = 0.155, e = 0.75, and CF = 0.25. High estimate assumes CC = $220/kWe, T = 20 yrs, FCR = 0.160, e = 0.50, and CF = 0.25.

[qqq]Low estimate assumes CC = $69/kWe, T = 30 yrs, FCR = 0.152, e = 0.85, and CF = 0.25. High estimate assumes CC = $285/kWe, T = 20 yrs, FCR = 0.160, e = 0.70, and CF = 0.25.

[rrr]Low estimate assumes CC = $500/kWe, T = 30 yrs, FCR = 0.152, e = 0.90, and CF = 0.25. High estimate assumes CC = $600/kWe, T = 20 yrs, FCR = 0.160, e = 0.80, and CF = 0.25.

averaged about 7 to 8 mills per kWhe (DOE 1979). Electrical storage costs were estimated to fall within a 6 to 150 mills per kWhe range, as compared to residential electricity, which currently costs about 30 mills per kWhe (DOE 1979). Of course, these comparisons are of limited significance owing to the fact that many of the storage technologies under consideration will not be operational for a decade or more, at which time conventional energy costs are likely to greatly exceed current values.

Of all the energy storage technologies, thermal storage is the least expensive. Between the two principal thermal approaches, sensible heat storage is presently less expensive than storage in phase-changing materials. However, in some cases the relative attractiveness might change with additional development. Thermal energy storage will be important for a broad spectrum of solar thermal applications ranging from low temperature space heating to thermal-electric power generation. Long-term, seasonal storage in underground aquifers or other earth formations looks particularly attractive from an economic standpoint.

Thermochemical storage is a potentially attractive approach to storing thermal energy owing to such attributes as high energy densities, long-term storage capabilities without energy losses or the need for insulation, and the flexibility made possible by the transportable nature of the reaction products. Hundreds of possible chemical reactions are under study, many of which look promising. While preliminary cost estimates are encouraging, chemical storage remains a largely untried concept with much developmental work to be done. Hydrogen systems, however, are somewhat more advanced than thermochemical storage in general, having been the object of more concentrated efforts. Although Table 5-1 suggests that hydrogen storage will be quite expensive, the computations in the table assume relatively continuous operation for all storage systems, and thereby obscure potential advantages held by systems (including hydrogen) with long-term storage capabilities. Therefore, the economics of hydrogen and other chemical storage techniques would be much more favorable for applications requiring storage over extended periods of time. Economic advantages may also be gained by energy transmission in the form of hydrogen gas or in chemical heatpipes.

Batteries, by virtue of their modular nature and lack of specific siting requirements, are the most flexible method of storing electricity and are adaptable in principle to a very wide range of applications, including the propulsion of vehicles. Their versatility and motive capabilities will make them an important energy storage mode. Owing to a short cycle life and low specific energy, lead-acid battery storage is expensive and relatively unattractive for large-scale future use. Its widespread use will be constrained in any event by the limited availability of lead. On the other hand, advanced batteries with higher energy densities potentially offer one of the least expensive methods of storing electricity. Numerous batteries in this category look promising for both electricity storage and vehicular propulsion.

Pumped hydroelectric storage is shown to be among the least expensive methods of storing large amounts of energy for relatively long periods of time. This technique also happens to be the most established of all the candidate systems. However, it is inflexible and only feasible on a large scale. Siting restrictions, the need for large amounts of water, and environmental considerations may constrain its expansion, although the introduction of underground pumped storage facilities may alleviate some of these problems. The use of existing hydroelectric facilities for storage offers an alternate means of expanding the available capacity.

Compressed air systems offer a large-scale storage potential with less siting constraints than pumped hydroelectric storage. However, the combined compressed air–gas turbine system is not a pure storage concept, requiring a suitable fuel (such as oil) for operation, whose energy is directly added to augment the stored energy. Consequently, its use will most likely be limited to an interim period. When fuel costs are added to storage costs, the conventional hybrid systems look relatively unattractive. On the other hand, adiabatic compressed air systems offer "pure" storage potential and need not depend on scarce fuels for operation. New machinery will have to be designed before these systems can be deployed, and consequently, the ultimate costs remain uncertain. However, preliminary estimates place the costs of adiabatic compressed air storage within the range of some of the other electrical storage techniques. Therefore, these systems warrant further research and development as major candidates for large-scale utility storage.

Advanced flywheels appear to hold considerable promise as an inexpensive means of storing mechanical or electrical energy for periods of time on the order of one hour. They are perhaps economical for periods up to ten hours. One possible approach is to combine the use of flywheels for short-term storage with another system having longer term storage capabilities. The modular nature of flywheels makes them similar to batteries in terms of their potential range of storage applications. In addition, their high power densities make them potentially attractive for use in electric vehicles, perhaps in conjunction with batteries. The flexibility, low cost potential, and favorable environmental characteristics should make flywheels serious competitors with batteries for a variety of applications. However, advanced flywheel technology is relatively immature, and additional development is required before the ultimate role of flywheels can be determined precisely.

Energy storage in superconducting magnets is the least advanced of all the storage methods considered here and its commercial feasibility is at best uncertain. Therefore, the cost estimates presented in Table 5-1 are highly uncertain and consequently should not be given much weight. It can be predicted with some degree of confidence, however, that this technology will only be feasible on a large scale, is likely to be expensive, and probably cannot be introduced before the year 2000. Owing to their potential for high electrical storage efficiencies, magnetic systems are worthy of serious consideration. Major technological advances are required, however, before the practicality of this approach can be firmly established.

ELECTRIC TRANSMISSION

There are several steps in the transmission of electricity from the power plant to the ultimate load point. Voltages are initially stepped up for long distance transmission and then reduced in stages to the various levels needed for residential, commercial, and industrial consumption.

With few exceptions, long distance power transmission takes place in overhead lines carrying alternating current (AC) at the standard frequency of 60 hertz (hz) or 60 cycles per second. The trend in AC power transmission has been toward a continuous increase in transmission line voltage. Maximum voltages have increased from 287 kilovolts (kV) before 1953 to 345 kV, 500 kV, and up to the present limit of 765 kV. As of 1970, there were approximately 300,000 miles of transmission lines installed with voltages exceeding 69 kV (Hodges 1976). First introduced in 1969, there are presently approximately 1,400 miles of 765 kV lines in operation (Miller and Kaufman 1978), with up to 10,000 or more miles planned by the year 1990 (Young and Young 1974). Utilities envision future lines up to 1,500 kV (Marino and Becker 1978), and facilities are already in existence to test transmission lines at voltages up to 2,200 kV (Miller and Kaufman 1978). Transmission at higher voltages is attractive because both power losses and unit power transmission costs are reduced. However, the economic advantages may be at least partially offset by possible hidden costs in the form of increased health and environmental risks, a matter we discuss later in this chapter.

Before utilization by the consumer, the electricity delivered by high capacity transmission lines is stepped down in voltage, with the reduction normally occurring in two stages. At bulk power substations, voltages are initially stepped down to between 4 and 35 kV. At local substations, levels are further reduced to 220 and 110 volts for residential and commercial usage. "Distribution" refers to that part of the electric power system that takes electricity from the bulk power substation to the point of end use. Both overhead and underground electric lines are used in distribution, with the latter employed predominately in congested, urban areas. Due to their presently adverse economics, underground lines account for only 1 percent of the current total transmission and distribution mileage (Kimbark 1976).

Technology

High voltage, direct current (DC) transmission offers an alternative to AC transmission. It is presently under development. The first major high voltage DC installation in the United States is an 800 kV line running between the Pacific Northwest and Southern California. DC lines have advantages and disadvantages in comparison to AC ones. The advantages are lower electrical losses, reduced insulation requirements, and the smaller cable size needed to transmit a given amount of power. It has been estimated that DC electric transmission can move energy thousands of miles with the loss of no more than about 5 percent (Caputo 1977). The greatest disadvantage is that voltage changes are less easily and efficiently obtained than with AC.

At present, underground cables are used in this country almost exclusively for distribution in urban areas, although long distance applications are under study. The most widely used underground transmission line consists of steel pipe enclosing three conductors, each wrapped in oil-impregnated paper for insulation. Voltages up to 500 kV have been carried by this kind of cable. However, high electrical losses limit this approach to short distances. In a method widely used in Europe, significant reduction in power losses can be obtained through the use of gas-insulated cables. In the most prevalent system of this type, sulfur hexafluoride (SF_6) is used both for insulation and as a coolant.

Cryoresistive cables take advantage of the fact that electrical resistivity decreases as the temperature is lowered. Low temperatures can be maintained by the circulation of liquid nitrogen, at a temperature of about minus 320°F. At this temperature, the resistance of the conductor is reduced by a factor of roughly ten compared to ambient temperature (Hammond et al. 1973). Because electrical losses are lower, such lines can handle higher power densities.

Superconducting transmission systems operate at even lower temperatures and are based on the principle that below a specific temperature, certain metals lose all electrical resistivity. The major advantage offered by superconducting power transmission are the elimination of resistive electrical losses, a reduction in cable material, and the flexibility of operation at high currents and reduced voltage. By operating at lower voltages, dielectric losses in AC cables can be considerably reduced. DC is probably more suitable for long distance superconducting lines and offers the ultimate in lossless transmission. Another advantage is the compact nature of superconducting, DC cables. A single cable, twenty-four inches in diameter, could carry more than 10,000 MW, exceeding New York City's peak power requirement of about 8,000 MW (Schwartz and Foner 1977).

While electrical losses can be nearly eliminated in superconducting cables, the power required for refrigeration is not negligible. In addition to refrigeration equipment, elaborate thermal insulation is also required. On balance, however, it appears that the power needed to operate the superconducting systems is likely to be less than the power losses and cooling requirements of all other transmission options (Haid 1976). Nonetheless, superconducting power transmission is the least advanced of the technologies we have discussed, and a major technical effort will be needed before commercialization can be seriously contemplated.

Usually many generating stations are interconnected, so that power can be supplied to users in the event of scheduled or unscheduled plant shutdowns. Many tens of millions of persons may be served by an interconnected system. Large national and international power grids are commonplace in Western Europe. In this country, there are regional power pools, but a national power grid is not yet quite a reality. In 1976, a 100 MW transmission line was installed that linked the eastern and western power grids (Stuart 1976), but this link will permit only small exchanges of power between the two systems.

Although the grid presently in place is a substantial start, several improvements are possible. According to a report by the General Accounting Office (1977), there is a need to construct additional utility interconnection links and to strengthen existing links by the installation of higher capacity transmission lines.

A scarcity of strong interconnections and weak coordination of diverse power systems was responsible for the northeastern power blackout of November 9, 1965, involving a total of about thirty million people in New York, New England, and parts of Canada. After this experience, it was obvious to many investigators and systems engineers that the Consolidated Edison system needed better fail-safe equipment and many more high capacity interconnections.

Improvements made following 1965, however, were not sufficient to avert a subsequent New York City blackout on July 13, 1977. Although inadequacies in the internal protection systems were partly at fault, a scarcity of grid interconnections lay at the heart of the difficulties that culminated in the blackout (Metz 1977). A chaotic period of twenty-six hours lapsed before the restoration of full power. Had more high capacity ties to outside electric grids been in place, the entire mishap might have been avoided.

On balance, a tightly interconnected national grid appears to offer, at least in theory, benefits of improved reliability. Most power companies maintain generators (referred to as "spinning reserve") that are ready to produce electricity but are not connected into the system. By tying individual systems with their reserves into a centrally dispatched grid, all the reserves could be available to meet unusually high loads of one or more members of the system. If backup high voltage lines are available, power can be sent to a system that has lost power through the breakdown of a generator or a transmission line. Electrical relays can automatically decouple a local utility from the national system in the event of failure. In fact, measures taken to speed the emergency disconnection from the grid were successful in confining the July 1977 power failure to New York City and Westchester County, avoiding the widespread regional blackouts occurring throughout the Northeast in 1965.

While strongly interconnected grids appear to offer generally enhanced reliability, the consequences of failure can be especially severe owing to the increased complexity of the system. Some observers feel that small-scale, autonomous electrical systems are less fragile and less susceptible to catastrophic failure than large, complex, and highly centralized systems (Lovins 1977).

Other potential advantages of well-integrated grid systems, however, are less controversial than the question of overall reliability. Transfers of electricity between interconnected power pools can reduce the requirements for standby peaking generating capacity, yielding economic benefits. In a linked, centrally controlled system, the relatively expensive inefficient peaking power plants can be relied on to a lesser extent, resulting in both dollar and energy savings. For example, in countries such as England and Sweden, the most efficient plants throughout the grid are brought on line first, with others added in order of decreasing efficiency (Wicklein 1977). In the

United States, on the other hand, where interconnections are less extensive, the less efficient and costlier plants have to operate a greater fraction of the time.

In the long term, an optimized national grid might be composed of integrated systems existing on several scales. Individual utilities would feed power into the grid and draw power from it. Local systems would be interconnected in a centrally controlled regional grid. The regional grids would in turn be linked to a national system that could transfer power from region to region to cover the peak or emergency demands of a particular region. Fail-safe measures would be made available to confine the effects of power failures to a localized area in the event of an emergency.

Economics

Over the past decade, installation, operation, and maintenance costs of electric transmission and distribution have accounted for roughly 40 percent of the total cost of producing and delivering electricity (*Electrical World* 1977). The cost of transmission systems will continue to be an important factor in the future, although some investigators have predicted significant cost reductions by the year 2000 for overhead and underground transmission (Davitian 1974).

Capital cost estimates for an overhead, 765 kV AC line are on the order of $180 per MW mile (OTA 1978). In terms of energy transmitted over 1,000 miles, costs cited for 500 kV and 765 kV AC overhead lines are about $0.013 (13 mills) per kWh (Konopka and Wurm 1974) and $0.010 (10 mills) per kWh (OTA 1978), respectively.

It has been estimated that the cost of an overhead DC line is currently about two-thirds that of a corresponding AC line (Kimbark 1976). Because voltage changes can be more easily achieved with AC electricity, rectifiers and inverters must be employed to convert from AC to DC and vice versa. This equipment is presently expensive, but advances in solid state technology could reduce costs substantially. Overhead DC lines are expected to be competitive with AC for the transmission of electricity over distances exceeding 400 to 500 miles (Hingorani 1978). The potential cost of high voltage overhead DC electric transmission over a distance of 1,000 miles has been estimated to be about 0.5 cents (5 mills) per kWh (Caputo 1977), or roughly half the cost cited for a comparable AC transmission line.

Underground transmission cables presently cost from six to twenty times more than overhead cables of similar capacity (Hammond et al. 1973; Kimbark 1976), whereas underground distribution is only two to three times as expensive as comparable overhead capacity (Baughman and Bottaro 1976). Underground transmission may have a cost advantage when land costs are considered, however. For example, land costs alone for overhead transmission installations are likely to exceed the cost of underground cables in densely populated areas (Hammond et al. 1973). Underground DC transmission is considered to be competitive with underground AC transmission at distances of about twenty-five miles or greater (Hingorani 1978). Though accurate cost predictions are difficult to make given the present state of the art, preliminary studies indicate that long distance electricity transmission via underground cryoresistive or superconducting cables may be economically competitive with other options for the transmission of bulk power rated at 2,000 MW or greater (Hein 1974; Lawrence and Cronin 1977).

Environmental and Health Considerations

The U.S. electric utility industry is one of the nation's greatest land users. The 300,000 miles of installed overhead transmission lines alone require four million acres (CEQ 1973). If power line rights of way are widened for reasons discussed later, this exclusion area would be further expanded. In addition to the problems of land use and visual pollution, there are other effects of high voltage overhead transmission. For example, television and radio interference can be troublesome, particularly in rural areas where higher levels of background noise are permitted. Noise pollution can also be a problem. Noise levels up to seventy decibels—approaching the legal long-term exposure limit for U.S. workers—have been recorded in the vicinity of 765 kV lines (Young 1973).

The potential health effects of the electromagnetic fields emanating from high voltage transmission lines, however, represent a subtler and relatively poorly understood phenomenon and consequently are much more controversial. While there is a growing body of scientific literature identifying biological effects of electromagnetic fields on humans and animals, neither the precise mechanism of interaction nor the degree of risk posed specifically by transmission lines are yet known. Electric charges accumulated on ungrounded metal surfaces in the vicinity of high voltage lines (such as tractors) can deliver a pronounced electric shock to a person upon contact. The attendant flow of electrical current through human subjects could have some adverse consequences (Young and Young 1974; Marino and Becker 1978). Evidence suggests that exposure to electric fields can affect growth and the central nervous and cardiovascular systems (Becker 1977). The effects of acute exposure to magnetic fields are generally considered to involve alterations in behavior and perception (Friedman et al. 1976; Marino 1977a).

Although a number of investigators have examined the biological effects of electric or magnetic fields, relatively few experiments have been designed to determine the effects of the combined electric and magnetic fields from transmission lines. The most extensive studies of the health effects of transmission line fields were carried out in the Soviet Union during the 1960s and 1970s. In one of these studies, an examination of workers at 400 and 500 kV substations revealed that forty-one out of forty-five experienced some cardiovascular or neurological disorder during and shortly after field exposure (Hill 1975). Adverse health effects have been cited in several other Soviet studies (Marino 1977a; Young and Young 1974; Young 1978). No health effects were observed, however, in a study funded by an American electric utility company examining a much smaller group of U.S. workers—ten high voltage linemen—carried out at Johns Hopkins University from 1962 to 1972 (Singewald et al. 1973).

As a result of the studies carried out in the Soviet Union, stringent exposure limits for Soviet power plant personnel were set in 1971 (Young and Young 1974). The exclusion area surrounding new 750 kV lines in the Soviet Union is considerably greater than that provided by U.S. standards (Marino 1976). Concluding that chronic exposure to 765 kV transmission lines would probably cause biological effects in humans, the New York Public Service Commission decided in 1977 that the protective zone around these lines should be extended from the present value of 250 feet to a minimum of 350 feet and to a maximum of 1,200 feet in order to ensure adequate protection (Faber 1977; Marino and Becker 1978).

An entirely different set of environmental problems might be associated with the phenomenon of corona discharge, which involves the actual emission of electrons from the conductor to the atmosphere. Corona discharge takes place when the voltage reaches a level high enough that air loses its effectiveness as an insulator. Emitted electrons may interact with molecules in the air to trigger the chemical formation of pollutants (Young 1973). It is not yet known whether this effect actually constitutes a problem of any real importance, but available evidence suggests that it is a comparatively trivial concern (Marino and Becker 1978).

Radiation from electrical power transmission lines has been observed to disturb the earth's magnetosphere (Helliwell et al. 1975; Stiles and Helliwell 1977; Park and Helliwell 1978), causing some degree of enhanced ionization, heat, optical emissions, and x-rays in the upper atmosphere. The x-rays produced could ultimately result in increased ultraviolet radiation reaching the earth's surface

(Marino and Becker 1978). However, whether these effects are of significance in comparison to natural geophysical processes has yet to be determined.

Overall Assessment

With the eventual exhaustion of abundant supplies of convenient fuel forms such as oil and natural gas, we believe that electricity will assume an increasingly important role in the future. However, uncertainties exist concerning both the ultimate cost and the associated health and environmental risks of the various transmission options.

Additional transmission lines will be needed in the short-term, and significant benefits, in terms of improved reliability, can be provided by bolstering weak links within the present utility grid system. The potential risks of overhead transmission can be diminished by the installation of overhead lines rated below 765 kV, until the hazards are better understood. Serious consideration should also be given to the expansion of rights of way surrounding future high voltage installations. If necessary, high voltage AC lines can be built—albeit at a higher cost—with reduced corona discharge by employing larger conductors with wider separation distances (Young 1973). The costs of retrofitting existing lines to minimize health risks appear to be substantial (Marino and Becker 1978). Long distance transmission by DC lines offers potential advantages in terms of improved efficiency and reduced costs.

Placing the lines underground is a longer range measure that is the most effective way of minimizing the environmental and health effects of power transmission. Problems relating to land use, visual pollution, television and radio interference, noise pollution, and corona discharge would all be eliminated. According to Becker (1977) and Marino (1977b), shielding from the earth could reduce ground level electromagnetic fields to ambient levels so that any possible adverse biological effects would be avoided. Utility corridors providing multiple rights of way for transmission lines, pipelines, and telephone lines would further minimize environmental disruption and also improve the economics of underground transmission. Of all the competing underground transmission technologies, superconducting DC cables appear to be the most attractive in terms of reduced volume and high efficiency. However, additional developmental work will be needed to firmly establish the viability of this approach. In any event, the most crucial consideration is to install the lines underground; all other considerations appear to be of secondary importance.

In the long term, it is possible to envision a fully integrated, national power grid transmitting electricity in highly efficient, underground cables. This network could conceivably link geographically dispersed solar and wind electric-generating facilities, interconnecting power pools of progressively larger scales. Such a system could provide reliable electric power with minimal assault on the environment and consequently warrants serious consideration as a major national goal.

HYDROGEN

Introduction

Hydrogen currently plays an important role in the fertilizer, chemical, food-processing, petrochemical, and metallurgical industries. The largest single use of hydrogen is in the synthesis of ammonia for fertilizers. Large quantities are also used in the U.S. space program as a rocket fuel in conjunction with oxygen or fluorine. Hydrogen may play a particularly vital role in transportation and energy storage and transmission applications in a renewable energy future. We concentrate on these potential future uses in this discussion and neglect current applications.

Hydrogen as a fuel has many inherent advantages. It has the highest energy content per unit mass of any chemical fuel and may be substituted for hydrocarbons in a broad range of applications, often with increased combustion efficiency. It can be utilized in fuel cells to produce both electricity and usable heat or burned in highly efficient turbine power systems. Hydrogen is in many ways an attractive transport fuel, especially for aircraft. Its combustion yields water as the major by-product, without the presence of carbon monoxide, carbon dioxide, or unburned hydrocarbons. Formation of nitrogen oxides can also be greatly reduced relative to those released by hydrocarbon combustion.

Drawbacks exist as well. Because of its lower densities, the energy content per unit volume of both gaseous and liquid hydrogen is only about one-third that of corresponding gaseous or liquid hydrocarbon fuels. It is a highly volatile substance whose handling requires extreme vigilance. Because of an extremely low boiling point (−424°F), its handling in liquid form necessarily involves the complexities of cryogenic operation. Transmission in existing natural gas pipelines may possibly be ruled out for a number of reasons, necessitating the construction of a new pipeline network.

The overall role that hydrogen will assume in the future is presently uncertain. It appears to be the most attractive fuel alternative for future use in aircraft and is also likely to find some use in water transportation. Safety considerations may favor the use of electricity over hydrogen as a means of powering the bulk of the future land transportation sector. Numerous possibilities exist for the use of hydrogen in fuel cells and for the performance of various thermal energy-requiring tasks. However, the extent to which hydrogen will be used for these applications cannot be confidently predicted until major economic uncertainties pertaining to the costs of production, storage, transmission, and conversion relative to electrical systems are accurately established.

Technology

Hydrogen energy systems may be broken down into the following components—production, storage, transmission, and utilization.

Hydrogen Production. Most of the hydrogen used today is produced from the steam reformation or partial oxidation of hydrocarbons such as natural gas. The production of hydrogen from coal is presently under development.

Among the alternative methods proposed for the generation of hydrogen, the electrolysis of water is the only commercially available technology. The efficiency of converting electricity into the chemical energy stored in hydrogen is presently about 50–80 percent in industrial applications (EPRI 1976). However, by operating at a high pressure and temperature, efficiencies exceeding 90 percent have been achieved in electrolytic cells (Ramakur 1976). Because thermal energy absorbed from the surroundings may be utilized in the process, an efficiency of about 100 percent is considered to be a realistic target (Kalhammer and Schneider 1976). Even if a 100 percent efficient electrolysis process were developed, the overall efficiency of hydrogen production would still be limited to that of electrical generation, which for electricity produced from primary fuels by a steam cycle is about 33 percent.

The need for electricity in the generation of hydrogen can be eliminated by using thermal energy to chemically split water. The direct thermal decomposition of water requires temperatures in excess of 4,000°F in order for meaningful efficiencies to be achieved (Nakamura 1977). A major barrier to this technique is the present lack of effective methods to separate hydrogen and oxygen at these high temperatures. Fletcher and Moen (1977) have proposed a one step, high temperature process with a potentially high efficiency that facilitates the separation of hydrogen and oxygen by means of effusion through a membrane. Efficiencies of about 80 percent have been projected for a similar process under development in Japan (SED

1977a). While concentrating solar collectors are capable of delivering thermal energy at the required temperatures for the direct splitting of water, numerous materials-related problems need to be resolved before this approach can be considered practical.

Water can be decomposed at a lower temperature in multistep thermochemical reactions in which thermal energy is used to catalyze the chemical reduction of water, rather than directly breaking up the molecular bonds. The highest temperatures required for the majority of these reactions range from about 930°F (Wentorf and Hanneman 1974) to about 3,100°F, with the average limit occurring near 1,500°F (JPL 1975). Theoretical efficiencies for these reactions range from 15-60 percent (Eisenstadt and Cox 1975; Wentorf and Hanneman 1974; Maugh 1972), although actual values remain to be determined in practice. Thermochemical generation is presently in the research stage, and several technical problems relating to reaction kinetics, thermal efficiency, and corrosion need resolution. Breakthroughs might make this approach an attractive alternative to electrolysis.

Several other production techniques are under investigation and warrant brief mention. One method under study at the University of North Carolina uses organic ruthenium complexes to catalyze the chemical splitting of water by solar energy (C&EN 1976a). Estimated overall efficiencies for this process range from 10 to 50 percent. A similar concept being developed at the California Institute of Technology evolves hydrogen when a rhodium complex is dissolved in water and exposed to sunlight (SED 1977b; *Technology Review* 1978). However, the efficiency of this cycle needs dramatic improvement. Another proposal involves the production of hydrogen gas from the reaction of organic material with high temperature, solar-generated steam (Antal 1976). The overall efficiency of this process is estimated to be 70 percent. The reliance on a large biomass source may prove to be the biggest drawback of this approach. Most other techniques, including photoelectric, photolytic, and biological production, are currently impractical owing to extremely low efficiencies (JPL 1975; Ramakur 1976).

Hydrogen Storage.   The three basic methods of storing hydrogen are as a high pressure gas, as a cryogenic liquid, and in metal hydrides—each having certain advantages for specific applications. For large-scale utility applications, compressed storage appears to be the most attractive approach where suitable reservoirs, such as depleted oil or gas wells, can be found.

The storage of hydrogen in liquid form is likely to be used in conjunction with air and water transportation vehicles. For cryogenic storage, vacuum-insulated vessels are needed to maintain the temperature of hydrogen below -423°F. Cryogenic storage is attractive mainly because the volume of stored hydrogen is reduced by a factor of about 850 from gas storage at 1 atm pressure (EPRI 1976). The storage and handling of liquid hydrogen poses a flammability hazard, however. In addition, the liquefaction of hydrogen is an energy-intensive process, requiring an amount of energy equivalent to about 25–30 percent of the heating value of the stored hydrogen (Gregory and Pangborn 1976).

Metal hydride storage can be achieved by exposing suitable metals or alloys to pressurized hydrogen. The gaseous hydrogen penetrates the metal and forms chemical bonds within the lattice structure. The density of hydrogen stored per unit volume is quite high and, for many hydride systems, is comparable to that of liquid hydrogen storage. Hydride formation is an exothermic process, and consequently, energy (called the heat of dissociation) is needed to drive the hydrogen-liberating process. The amount of energy required for dissociation represents about 10 percent of the energy content of the hydrogen stored in iron-titanium hydrides, 31 percent for magnesium hydrides, and up to 75 percent for some systems (JPL 1975). The iron-titanium hydride is favored for storage because of its low heat of formation; overall storage efficiencies of 95 percent have been measured for this system (EPRI 1976). The iron-titanium hydride is very heavy in comparison to other hydrides, however, and therefore would not be suitable for mobile applications. Magnesium hydrides, holding more hydrogen per unit weight, appear to be the most attractive for use in vehicles.

Hydrogen Transmission.   The most attractive method of hydrogen transmission is almost certainly to be by gas pipelines, although hydrogen may also be transported in liquid and hydride form by rail, truck, or freighter. There has been some limited experience with the operation of hydrogen pipelines. A twelve mile long pipeline in the United States has been operated routinely without any specific precautions having been taken. The most extensive hydrogen pipeline network is in West Germany. This system is 130 miles long and has been operated for nearly forty years without any major incidents (Gregory and Pangborn 1976).

A number of factors make the transmission of hydrogen different from natural gas and could necessitate the construction of new pipelines. The lower volumetric energy density of hydrogen (approximately 325 Btu/ft^3) in comparison to natural gas (about 1,000 Btu/ft^3) means that the volume of hydrogen handled by a pipeline would have to exceed that of natural gas by a factor of more than three in order to transmit an equivalent amount of energy. Therefore, compressor energy requirements would be increased for hydrogen, resulting in added costs. Although existing natural gas pipelines conceivably could be used to handle hydrogen under low pressure, transmission would be inefficient (Bockris 1975). The number of pumping stations would have to be increased, and in addition, joints would have to be carefully examined in order to guard against leaks. For the optimal transmission of hydrogen, new pipelines—having different diameters and modified compressors—would need to be designed.

Another matter that would affect the transmission of hydrogen in pipelines, as well as affecting other links in hydrogen energy systems, is the materials problems caused by hydrogen embrittlement. The term embrittlement refers to hydrogen's ability to penetrate certain sensitive metals, such as low alloy steel, and weaken them, possible leading to corrosion and breakdown. Gregory and Pangborn (1976) concluded that while more research is needed, present evidence indicates that the embrittlement experienced by normal pipeline steel is likely to be minimal for operating pressures up to 750 psia. For higher pressures, there is much less certainty about the integrity of these conventional materials. A new high performance stainless steel that is suitable for hydrogen pipeline applications has been developed at Sandia Laboratories in Livermore, California, but production cost estimates are unavailable (SED 1977c). One question that remains unanswered is whether suitable materials can be found at reasonable costs.

Hydrogen Utilization. Hydrogen can serve as an intermediate energy form in the storage of both electrical and thermal energy. Hydrogen generated by electrolysis can be stored and then pipelined to its ultimate destination. Hydrogen may be reconverted to electricity in fuel cells at a maximum efficiency of about 70 percent or in advanced hydrogen-oxygen gas turbines with efficiencies of 50–55 percent (JPL 1975). Fuel cells are nonpolluting conversion devices that can be located at the load site and are capable of providing both electrical and thermal energy. Rechargeable fuel cells store electricity in a manner similar to batteries, without hydrogen ever leaving the storage unit.

On the basis of rather optimistic assumptions, an upper range

Photograph 5-2: Fuel Cell. Schematic of a 40 kilowatt fuel cell designed to run on air and natural gas. The unit is being developed by United Technologies Corporation of Hartford, Connecticut. Advanced fuel cells may use hydrogen as a fuel.

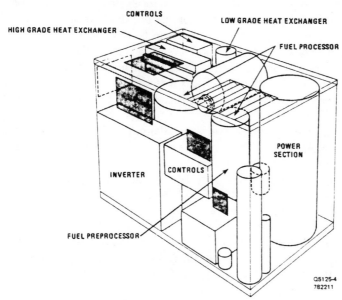

Source: United Technologies Corporation.

estimate for the electrical energy storage efficiency of hydrogen systems would at most be about 63 percent.[7] A comparison with the corresponding electrical storage efficiencies presented in Chapter 6 reveals that even under the most optimistic conditions, storage efficiencies for hydrogen would fall near the lower estimated range for the competing technologies. Storage efficiencies appear more favorable when hydrogen is produced from thermal energy, if reasonably efficient production methods can be established. If the hydrogen so produced were stored and subsequently used to generate electricity in fuel cells, the overall thermal efficiencies might be comparable, if not superior, to on-site thermal energy storage and electrical generation.[8] Therefore, hydrogen generation and storage could conceivably

7. Assumes 100 percent efficient electrolysis, 90 percent storage efficiencies in metal hydrides, and 70 percent efficient fuel cells.
8. Assuming a thermochemical production efficiency of 50 percent, hydride storage efficiencies of 90 percent, and 70 percent efficient fuel cells, as compared to 80 percent thermal storage efficiencies and 33 percent electrical-generating efficiencies.

provide an attractive option for load leveling in a thermal power plant.

The use of hydrogen as a fuel for land, water, and air transportation has been frequently suggested. Of the three, applications in the air transportation sector appear to be the most promising. Hydrogen is an attractive fuel because of its environmental advantages and high energy content per unit mass in comparison to conventional jet fuel—51,600 Btu/lb for hydrogen (lower heating value) as opposed to 18,600 Btu/lb for the latter. The higher specific energy of hydrogen results in a lighter fuel load and overall weight savings contributing to improved performance. With relatively simple modifications, aircraft gas turbine engines can be made to operate on hydrogen—a fact demonstrated in the 1950s (Gregory and Pangborn 1976). The weight advantages stemming from the utilization of hydrogen become especially important in high speed supersonic and hypersonic air travel. Liquid hydrogen is in fact the only fuel suitable for hypersonic aircraft, because apart from weight considerations, it could be used to provide the cooling necessary to prevent the craft's surfaces from overheating.

Liquid hydrogen may also prove to be a convenient fuel for marine vessels. It would probably be best suited for shorter range domestic shipping, for the reason that the long storage times imposed in international shipping would require extremely effective insulation to avoid excessive hydrogen boiloff. One study found that the use of hydrogen in place of conventional diesel fuel would result in either equivalent or improved performance and range depending on the type of vessel (JPL 1975).

The widespread use of hydrogen as a fuel for automobiles, trucks, and buses faces formidable barriers despite hydrogen's clean-burning characteristics and superior efficiency in comparison to gasoline. The on-board storage of hydrogen constitutes the major problem. Storage as a compressed gas would generally appear to be impractical because the container weight would be excessive—with the equivalent of a twenty gallon gasoline tank weighing about 2,250 pounds (JPL 1975). However, a high pressure cylinder has been developed in Germany that is capable of storing enough hydrogen gas to drive a Volkswagon 200 miles (SED 1977d). The storage of liquid hydrogen would require an expensive tank with elaborate insulation and would involve serious safety hazards in fuel handling and from tank boiloff. Metal hydrides offer the most practical approach, with the major shortcomings being their weight and the need to apply thermal energy for dissociation. One U.S. company has installed several modified automobiles and buses with hydride tanks enabling them to utilize hydrogen fuel (SED 1977e). A magnesium hydride presently appears to be the most attractive for vehicles. The weight of a system with a hydrogen-carrying capacity equivalent to twenty gallons of gasoline would be about 700 lb; the dissociation energy would be equivalent to about 31 percent of the energy content of the stored hydrogen (JPL 1975). The dissociation energy requirement might be provided, at least in part, by the utilization of engine exhaust heat, although this concept has yet to be demonstrated. Regardless of the storage method selected, safety questions relating to the dispensation and storage of hydrogen would require a thorough and critical evaluation.

The major limiting factor may be that electricity will ultimately prove to be more suitable than hydrogen for the task of powering a land transportation sector (see Chapter 6). Hydrogen does not appear to have any decisive advantage on efficiency grounds, and electricity may prove to be superior from the standpoints of safety, cost, and convenience. Further research and development is needed to determine conclusively the trade-offs between the two approaches.

Economics

While some of the hydrogen subsystems are in the commercial stage, others are at a developmental level, so that the ultimate costs of hydrogen systems are uncertain, although they appear likely to be high. Overall capital costs for "intermediate-term" hydrogen systems consisting of an electrolyzer, compressed gas storage (ten hours capacity), and fuel cells have been estimated at about $600–1,100 per kWe (EPRI 1976). A more optimistic figure of $300 per kWe has been projected for future electrolyzer-hydride-fuel cell systems (Kalhammer and Schneider 1976). A comparison with the capital costs of other storage technologies presented earlier in this chapter indicates that hydrogen systems will be among the more expensive. However, the ostensibly high costs may be at least partially offset by possible cost advantages for transmission and distribution or by breakthroughs in thermal production methods that would obviate electrical generation. In any event, certain applications exist for which hydrogen is likely to be the fuel of choice, regardless of its precise cost.

Estimates for the cost of hydrogen fuel—produced by electrolytic or by thermochemical techniques from solar, wind, and conventional sources—vary widely depending on the assumptions made, but generally fall within the range of $1–11 per million Btu (Ramakur 1976; Bockris 1975; Eisenstadt and Cox 1975; Heronemus 1976). A probable value of $5–6 per million Btu has been suggested by Bockris (1978). This is approximately twice the 1978 cost of fuels such as natural gas, heating oil, and gasoline, which averaged roughly $2-3 per million Btu (DOE 1979).

Although hydrogen transmission may prove to be less expensive than electricity transmission, it is difficult to claim an assured advantage for either approach at the present time. The cost of hydrogen transmission in a completely optimized pipeline system has been estimated to be about $0.035-0.055 per million Btu per hundred miles, higher than present transmission costs for natural gas, but lower than the $0.06-0.37 per million Btu per one hundred miles estimate made for electric transmission in a high voltage overhead line (Konopka and Wurm 1974; Davitian 1974). However, most estimates for hydrogen transmission assume that normal pipeline steels will prove to be suitable. If embrittlement problems force the use of higher strength and costlier steel, the relative economics will be altered to an uncertain extent. Estimates made by Bockris (1978) place the cost of hydrogen transmission in the same range as those cited for electricity and also demonstrate the sensitivity of these costs to the total distance of transmission. He estimated a cost of $0.41 per million Btu per one hundred miles for a total transmission distance of one hundred miles and that of $0.062 per million Btu per one hundred miles for a pipeline a thousand miles in length, strongly suggesting that the use of hydrogen as an energy carrier will be most economical for applications involving long distance transmission.

Environmental and Safety Considerations

The environmental impacts associated directly with the use of hydrogen—either its combustion or its electrochemical combination with oxygen—are expected to be minimal. The principal combustion product is water, although some nitrogen oxides might be produced if the temperature is sufficiently high. Because nitrogen oxides are the only potential pollutant, they could be more easily contained than in a hydrocarbon combustion system (Gregory and Pangborn 1976). For example, the injection of water vapor into the engine combustion chamber in one hydrogen-powered bus kept combustion temperatures below the level needed for nitrogen oxide formation (C&EN 1976b). The catalytic combustion of hydrogen has also been demonstrated. It allows combustion to proceed at a temperature lower than in a conventional flame burner, so that nitrogen oxide production may be reduced to a negligible level (Gregory and Pangborn 1976). Thus, major benefits could be realized from the combustion of hydrogen in place of hydrocarbon fuels.

The major environmental impacts of hydrogen utilization would be associated with hydrogen production and the manufacturing of related equipment. The environmental effects of hydrogen will vary considerably depending on whether coal, nuclear energy, or solar energy is used as the primary source.

On balance, hydrogen is somewhat more difficult to handle than natural gas. Hydrogen forms a combustible mixture with air over a wider range of mixture ratios than natural gas. The ignition energy and ignition temperature are lower for hydrogen. At normal temperatures, hydrogen gas heats upon expansion, with the possibility of ignition. Therefore, the possibility of fuel-air explosions in confined spaces is greater for hydrogen. However, certain properties of hydrogen make it less dangerous than natural gas. Hydrogen contains less energy per unit volume and also diffuses more rapidly— factors that tend to mitigate its explosive potential in open areas. In addition, hydrogen fires are less hazardous than natural gas ones in terms of smoke inhalation risk.

While vigilance is required, hydrogen has been safely transported in pipelines in Europe and, to a more limited extent, in the United States. Safety problems, however, may be sufficient to prevent the general use of liquid hydrogen in automobiles and also rule out the widespread distribution and dispensation of hydrogen at refueling stations. On the other hand, liquid hydrogen has considerable potential as an aircraft fuel, in light of the safe handling of cryogenic hydrogen in the U.S. space program.

Overall Assessment

While electrolytic production of hydrogen is presently the most established method, thermal production techniques, which obviate the need for electrical generation, offer the greatest potential for increases in overall efficiency and lowered costs.

A mix of hydrogen storage methods is likely to be employed in the future. Underground compressed gas storage appears most suitable for large-scale utility applications. Cryogenic storage is most compatible with applications requiring liquid hydrogen, such as air and water transportation, whereas metal hydrides offer the most practical approach for land vehicles.

The lower volumetric energy density of hydrogen relative to natural gas will necessitate the construction of newly designed pipelines for optimal transmission. Embrittlement problems may require the use of new pipeline steels, whose costs are uncertain. Therefore, any potential transmission and distribution cost advantages that may be ascribed to hydrogen remain to be demonstrated.

Of all the potential uses of hydrogen, the most promising applications appear to be in the air and water transportation sectors. On the basis of overall safety and convenience, electricity may offer a more attractive approach to powering the land transportation sector. Many other possibilities exist for hydrogen, including the use in fuel cell-total energy systems and for the satisfaction of tasks requiring thermal energy. However, the ultimate role that hydrogen will assume cannot be clearly delineated until the many technical and economic uncertainties related to production, storage, transmission, and conversion vis-à-vis electricity are more accurately determined.

    *   *   *   *   *

Storage and transmission comprise two of the more neglected links in the energy chain. As supplies of easily stored fossil fuels become increasingly scarce, the need to develop and deploy new energy storage systems will grow commensurately. In fact, storage will ultimately assume an importance comparable to that of energy supply and should therefore receive a correspondingly high priority. While a fairly extensive electricity and natural gas transmission system is already in place in the United States, further improvements can be made that would yield substantial advantages. Moreover, the establishment of a hydrogen pipeline distribution system might eventually become an important priority.

Current Department of Energy research and development budgets allocate less than $100 million annually to storage and transmission systems, while energy supply technologies receive more than $3 billion per year (DOE 1979b). Clearly, a more vigorous federal research, development, and demonstration program is crucial. If a concerted effort is now launched to develop and introduce advanced storage and transmission systems as they are needed, the groundwork can be laid for a smooth transition to a renewable energy future.

REFERENCES

Allen, A.E. 1976. "Potential for Conventional and Underground Pumped Storage." Paper presented at IEEE/ASME/ASCE Joint Power Generation Conference, Buffalo, New York, September 19–23.

American Physical Society (APS). 1979. "Executive Summary of the American Physical Society Study Group on Solar Photoelectric Energy Conversion." New York, January.

Andrews, J. 1976. "Energy Storage Requirements Reduced in Coupled Wind-Solar Generating Systems." *Solar Energy* 18:73–74.

Antal, M.J. 1976. "Tower Power: Producing Fuels from Solar Energy." *Bulletin of the Atomic Scientists* 32, no. 5 (May):58–62.

Applied Physics Laboratory (APL). 1975. "Proposal to New York State Energy Research and Development Authority for Flywheel Electric Energy

Storage System." Laurel, Md.: Applied Physics Laboratory, Johns Hopkins University, November 10.

Asbury, J.G., and P.D. Muller. 1977. "Solar Energy and Electric Utilities: Should They Be Interfaced?" *Science* 195, no. 4277. (February 4).

Baughman, M.C., and D.J. Bottaro. 1976. "Electric Power Transmission and Distribution Systems: Costs and Their Allocation." IEEE. Transactions on Power Apparatus and Systems PAS-95 no. 3 (May-June):782-90.

Becker, R.O. 1977. "Common Record Hearing on the Health and Safety of 765-kV Transmission Lines." Prepared testimony before the State of New York Public Service Commission, Cases 26529 and 26559.

Behrin, E., et al. 1977. "Energy Storage Systems for Automobile Propulsion, Vol. I: Overview and Findings." Livermore: University of California, UCRL-52303, December.

Birk, J.R., and F.R. Kalhammer. 1976. "Secondary Batteries for Load Leveling." Palo Alto, Calif.: Electric Power Research Institute, March.

Blackstone, S. 1975. "Superflywheel: Hope for Energy." *New York Times*, November 30, p. 84.

Bockris, J. 1975. *Energy: The Solar-Hydrogen Alternative.* New York: Halsted Press—John Wiley and Sons.

———. 1978. "Projections on the Price of Hydrogen Fuel." *Energy Research* 2:9-17.

Bush, J.B., Jr., et al. 1975. "An Assessment of the Technical and Economic Feasibility of Compressed Air Energy Storage." In proceedings of the Workshop on Compressed Air Energy Storage System, ERDA-76-124, sponsored by the Energy Research and Development Administration and the Electric Power Research Institute, December 18-19, Airlie House, Virginia, pp. 14-72.

Caputo, R.S. 1977. "Solar Power Plants: Dark Horse in the Energy Stable." *Bulletin of Atomic Scientists* 33, no. 5 (May):47-56.

Chang, G. 1977. Mechanical Energy Storage Division of the Energy Research and Development Administration, Washington, D.C., personal communication, May 26.

Chemical and Engineering News (C&EN). 1976a. "Ruthenium Complexes Aid Hydrogen Process." 54, no. 23 (May 31):17-20.

———. 1976b. "Hydrogen-Powered Bus Serves Citizens of Provo, Utah." 54, no. 23 (May 31):17.

———. 1977. "Battery Development Makes Good Progress." 55, no. 8 (February 21):28-30.

———. 1979. "Study Finds Chemical Heat Pipes Competitive." 57, no. 5 (January 29):7.

Corso, R.A. 1976. "Private Sector Hydroelectric Development." Paper presented at IEEE/ASME/ASCE Joint Power Generation Conference, Buffalo, New York, September 19-23.

Council on Environmental Quality (CEQ). 1973. "Energy and the Environment: Electric Power." Washington, D.C., August.

Davison, R.R. 1975. "Long Term and Seasonal Storage Group Report." In Proceedings of the Workshop on Solar Energy Storage Subsystems for the Heating and Cooling of Buildings, NSF-RA-NA-75-041, sponsored by the

National Science Foundation and the Energy Research and Development Administration, Charlottesville, Virginia, April 16-18, pp. 170-71.

Davitian, H. 1974. "Energy Carriers in Space Conditioning and Automotive Applications: A Comparison of Hydrogen, Methane, Methanol, and Electricity." In 9th Intersociety Energy Conversion Engineering Conference Proceedings, San Francisco, California, August.

Department of Energy (DOE). 1979. "Monthly Energy Review." Washington, D.C.: Energy Information Administration, DOE/IEA-0035/2(79), February.

———. 1979b. "Budget Highlights," Department of Energy FY 1980 Budget to Congress, DOE/CR-0004. Washington, D.C.

Eisenstadt, M.M., and K.E. Cox. 1975. "Hydrogen Production from Solar Energy." *Solar Energy* 17:59-65.

Electric Power Research Institute and Energy Research and Development Administration (EPRI). 1976. "An Assessment of Energy Storage Systems Suitable for Use by Electric Utilities." EPRI EM-264, ERDA E (11-1)-2051. Prepared by the Public Service Electric and Gas Company, Newark. New Jersey, July.

Electrical World. 1977 "28th Annual Electrical Industry Forecast." September 15.

Energy Research and Development Administration (ERDA). 1975. "Economic and Technical Feasibility Study for Energy Storage Flywheels." ERDA 76-65, UC-94B. Prepared by Rockwell International Space Division, December.

Ervin, G. 1975. "Solar Heat Storage Based on Inorganic Chemical Reactions." In Proceedings of the Workshop on Solar Energy Storage Subsystems for the Heating and Cooling of Buildings, NSF-RA-N-75-041, Charlottesville, Virginia, April 16-18, pp. 91-95.

Faber, H., 1977. "Public Service Commission Finds Biological 'Effects' Caused by High-Voltage Conduits." *New York Times*, September 18.

Federal Energy Administration (FEA). 1974. "Project Independence." "Final Task Force Report—Solar Energy." Under direction of the National Science Foundation, Washington, D.C., November.

Federal Power Commission (FPC). 1971. "1970 National Power Survey." Pt. IV. Washington, D.C.

Fischer, H.C. 1975. "Annual Cycle Energy System (ACES) for Residential and Commercial Buildings." In Proceedings of the Workshop on Solar Energy Storage Subsystems for the Heating and Cooling of Buildings, NSF-RA-N-75-041, Charlottesville, Virginia, April 16-18, pp. 129-35.

Fletcher, E.A., and R.L. Moen. 1977. "Hydrogen and Oxygen From Water." *Science* 197 (September 9):1050-56.

Friedman, H., et al. 1976. "Effect of Magnetic Fields on Reaction Time Performance." *Nature* 213 (March 4):949-50.

Gelb, G.H., et al. 1972. "Performance Analysis of Heat Engine-Battery Hybrid Vehicles." In Proceedings of the 7th Intersociety Engineering Conference, American Chemical Society, Paper No. 729144.

General Accounting Office (GAO). 1977. "Problems in Planning and Constructing Transmission Lines Which Interconnect Utilities." Report No. PB-269614. Washington, D.C., June 9.

Glendenning, I. 1975. "Long Term Prospects for Compressed Air Energy

Storage." In Proceedings of the Workshop on Compressed Air Energy Storage System, ERDA-76-124, sponsored by the Energy Research and Development Administration and the Electric Power Research Institute, Airlie House, Virginia, December 18-19, pp. 358-89.

Gregory, D.P., and J.B. Pangborn. 1976. "Hydrogen Energy." *Annual Review of Energy* 1:279-310.

Green, R.M., et al. 1976. "High Temperature Thermal Energy Storage." In *Sharing the Sun!* Joint Conference of the American Section of the International Solar Energy Society and the Solar Energy Society of Canada, Inc., Winnepeg, August 15-20, vol. 8, pp. 4-47.

Grodzka, P.J. 1975. "Some Practical Aspects of Thermal Energy Storage." In Proceedings of the Workshop on Solar Energy Storage Subsystems for the Heating and Cooling of Buildings, NSF-RA-N-75-041, Charlottesville, Virginia, April 16-18, pp. 68-74.

Gross, S. 1976. "Review of Candidate Batteries for Electric Vehicles." *Energy Conversion* 15:95-111.

Gruen, D.M., and I. Sheft. 1975. "Metal Hydride Systems for Solar Energy Storage and Conversion." In Proceedings of the Workshop on Solar Energy Storage Subsystems for the Heating and Cooling of Buildings, NSF-RA-N-75-041, Charlottesville, Virginia, April 16-18, pp. 96-99.

Haid, D.A. 1976. "Power Transmission Via the Superconducting Cable." *Mechanical Engineering* (January):20-25.

Hammond, A.L., et al. 1973. *Energy and the Future.* Washington, D.C.: American Association for the Advancement of Science.

Hein, R.A. 1974. "Superconductivity: Large-Scale Applications." *Science* 185 (July 19):211-22.

Helliwell, R.A., et al. 1975. "VLF Line Radiation in the Earth's Magnetosphere and its Association with Power System Radiation." *Journal of Geophysical Research* 80, no. 31 (November 1):4249-58.

Herman, S., and J. Cannon. 1976. *Energy Futures.* New York: Inform, Inc.

Heronemus, W.E. 1976. "Oceanic Windpower." Prepared for a Conference at North Carolina State University. Amherst, Massachusetts, University of Massachusetts.

Hill, G. 1975. "Ultrahigh-Voltage Lines Studied as Possible Peril." *New York Times,* November 10.

Hingorani, N. 1978. "The Re-Emergence of DC in Modern Power Systems." *EPRI Journal,* June, pp. 6-13.

Hobson, M.J., et al. 1976. "Underground Compressed Air Storage Power-Parametric Analysis of Cycles and Cost for Various Concepts." Presented at the IEEE/ASME/ASCE Joint Power Generation Conference, Buffalo, New York, September 19-23.

Hodges, L. 1976. "Energy Transmission." In *McGraw-Hill Encyclopedia of Energy,* D.N. Lapedes, ed., pp. 255-56. New York: McGraw-Hill.

Jet Propulsion Laboratory (JPL). 1975. "Hydrogen Tomorrow—Demands and Technology Requirements." Report of the NASA Hydrogen Energy Systems Technology Study. Pasadena: California Institute of Technology, December.

Joy, P., and B. Shelpuk. 1976. "Solar Heating-Thermal Storage Feasibility."

ASME, 76-WA/HT-36. New York: American Society of Mechanical Engineers.

Justus, C.G. "Wind Energy Statistics for Large Arrays of Wind Turbines (New England and Central U.S. Regions)." A report prepared for the National Science Foundation. ERDA/NSF-00547/76/1. Washington, D.C., August.

Kauffman, K.W., and H.G. Lorsch. 1976. "Design and Costs of High-Temperature Thermal Storage Devices Using Salts or Alloys." ASME, 76-WA/HT-34. New York: American Society of Mechanical Engineers.

Kalhammer, F.R., and T.R. Schneider. 1976. "Energy Storage." *Annual Review of Energy* 1:311-44.

Kimbark, E.W. 1976. "Power Transmission Lines." In *McGraw-Hill Encyclopedia of Energy*, D.N. Lapedes, ed., pp. 695-700. New York: McGraw-Hill.

Konopka, A., and J. Wurm. 1974. "Transmission of Gaseous Hydrogen." 9th Intersociety Energy Conversion Engineering Conference Proceedings, San Francisco, California, August.

Kreid, D.K. 1976. "Technical and Economic Feasibility Analysis of the No-Fuel Compressed Air Energy Storage Concept." Prepared for the Energy Research and Development Administration. Richland, Wash.: Battelle Pacific Northwest Laboratories, BNWL-2065, UC94b, May.

———. 1977. Batelle Pacific Northwest Laboratory, personal communication, May 26.

Kreider, J., and F. Kreith. 1975. *Solar Heating and Cooling*, New York: McGraw-Hill.

Lawrence, R.F., and J.H. Cronin. 1977. "Trends in Electric Power Transmission." In *Energy Technology Handbook*, D.M. Considine, ed. New York: McGraw-Hill.

Lorsch, H.G., et al. 1975. "Thermal Energy Storage for Solar Heating and Off-Peak Air Conditioning." *Energy Conversion*, pp. 1-6.

Lovins, A.B. "Resilience in Energy Strategies." *New York Times*, July 24.

Margen, P. 1978. "Central Plants for Annual Heat Storage." *Solar Age* 3, no. 10 (October):22-26.

Marino, A.A. 1976. Rebuttal Testimony—Cases 26529 and 26559—Common Record Hearings on Health and Safety of 765kV Transmission Lines, submitted to the Public Service Commission, Albany, New York, November 24.

———. 1977a. "Biological Effects of Extremely Low Frequency Electric and Magnetic Fields." State of California, State Energy Resources Conservation and Development Commission, Docket No. 76-NOI-2, June 17.

———. 1977b. Veteran's Administration Hospital, Syracuse, New York, personal communication, December 15.

Marino, A.A., and R.O. Becker. 1978. "High Voltage Lines: Hazard at a Distance." *Environment* 20, no. 9 (November):6-15.

Maugh, T.H., II. 1972. "Hydrogen: Synthetic Fuel of the Future." *Science* 178 (November 24):849-52.

McCaull, J. 1976. "Storing the Sun." *Environment* 18, no. 5 (June):9-15.

Metz, W.D. 1977. "New York Blackout: Weak Links Tie Con Ed to Neighboring Utilities." *Science* 197 (July 29): 441-42.

———. 1978. "Energy Storage and Solar Power: An Exaggerated Problem." *Science* 200 (June 30):1471–73.

Miller, M.W., and G.E. Kaufman. 1978. "High Voltage Overhead." *Environment* 20, no. 1 (January-February): 6–36.

Nakamura, T. 1977. "Hydrogen Production from Water Utilizing Solar Heat at Higher Temperatures." *Solar Energy* 19:467–75.

National Academy of Science (NAS). 1976. "Criteria for Energy Storage R&D." Washington, D.C.

Oakes, R. 1977. Midwest Bunker Silo Co., Portland, Michigan, personal communication, June 16.

O'Connell, L.G., et al. 1975. "The Lithium-Water-Air-Battery: A New Concept for Automotive Propulsion." Livermore: Lawrence Livermore Laboratory, University of California, Report No. UCRL–51811, May.

Offenhartz, P.O. 1976. "Chemical Methods of Storing Thermal Energy." In *Sharing the Sun!* Joint Conference of the American Section of the International Solar Energy Society and the Solar Energy Society of Canada, Inc., Winnipeg, August 15–20, vol. 8, pp. 48–72.

Office of Technology Assessment (OTA). 1978. "Application of Solar Technology to Today's Energy Needs." Vol. 1. Washington, D.C.: Government Printing Office, June.

Park, C.G., and R.A. Helliwell. 1978. "Magnetospheric Effects of Power Line Radiation." *Science* 200 (May 19):727–30.

Pickering, E.E. 1975. "Residential Hot Water Solar Energy Storage." In Proceedings of the Workshop on Solar Energy Storage Subsystems for the Heating and Cooling of Buildings, NSF–RA–N–75–041, Charlottesville, Virginia, April 16–18, pp. 24–37.

Putnam, P.C. 1948. *Power from the Wind.* New York: Van Nostrand Reinhold Company.

Ramakur, R. 1976. "Solar Technology in the Seventies—An Assessment of Hydrogen as a Means to Store Solar Energy." In *Sharing the Sun!* Joint Conference of the American Section of the International Solar Energy Society and the Solar Energy Society of Canada, Inc., Winnipeg, August 15–20, vol. 8, pp. 163–74.

Riaz, M., et al. 1976. "High-Temperature Energy Storage in Native Rock." In *Sharing the Sun!* Joint Conference of the American Section of the International Solar Energy Society and the Solar Energy Society of Canada, Inc., Winnipeg, August 15–20, vol. 8, pp. 123–38.

Ricci, L.J. 1975. "Utilities Eye Large-Scale Energy Storage." *Chemical Engineering* (February 3):24–25.

Robinson, A.L 1974a. "Energy Storage (I): Using Electricity More Efficiently." *Science* 184 (May 17):785–87.

———. 1974b. "Energy Storage (II): Developing Advanced Technologies." *Science* 184 (May 24):884–87.

Schroder, J. 1974. "Thermal Energy Storage and Control." ASME, 74–WA/ Oct–1. New York: American Society of Mechanical Engineers, October.

Schwartz, B.B., and S. Foner. 1977. "Large Scale Applications of Super-conductivity." *Physics Today* 30, no. 7 (July):34–43.

Scott, F.M. 1977. "Underground Hydroelectric Pumped Storage: A Practical Option." *Energy* II, no. 4:20–22.

Simmons, J.A. 1976. "Reversible Oxidation of Metal Oxides for Thermal Energy Storage." In *Sharing the Sun!* Joint Conference of the American Section of the International Solar Energy Society and the Solar Energy Society of Canada, Inc., Winnipeg, August 15–20, vol. 8, pp. 219–25.

Singewald, M.L., et al. 1973. "Medical Followup Study of High Voltage Lineman Working in AC Electric Fields." *IEEE Transactions of Power Apparatus and Systems*, PAS-92, pp. 1302–05. New York: Institute of Electrical and Electronic Engineers.

Solar Energy Digest (SED). 1977a. "A Solar Hydrogen Production Breakthrough." 8, no. 5 (May):7.

———. 1977b. "Compound Uses Sunlight to Produce Hydrogen." 9, no. 5 (November):4.

———. 1977c. "Transportation of Hydrogen." 9, no. 2 (August):10.

———. 1977d. "Hydrogen Stored in Cylinders Can Now Drive a Car 200 Miles." 8, no. 2 (February):8.

———. 1977e. "Billings Starts Converting Datsuns to Run on Hydrogen Fuel." 9, no. 4 (October):4.

Solar Energy Intelligence Report (SEIR). 1977. "GE Claiming a Breakthrough in Phase-Change Salt Storage for Solar Heating." (December 26):301.

Sorenson, B. 1976. "Dependability of Wind Energy Generators with Short Term Energy Storage." *Science* 194:935–37.

Stanford Research Institute (SRI). 1976. "Long Term Energy Alternatives for Automotive Propulsion-Synthetic Fuel Versus Battery/Electric System." Prepared for the National Science Foundation. Menlo Park, Calif.: PB-262 513, August.

Stephens, T. 1975. "Adiabatic Compressed Air Energy Storage Systems." In Proceedings of the Workshop on Compressed Air Energy Storage System, Sponsored by the Energy Research and Development Administration and the Electric Power Research Institute, Airlie House, Virginia, December 18–19.

Stiles, G.S., and R.A. Helliwell. 1977. "Stimulated Growth of Coherent VLF Waves in the Atmosphere." *Journal of Geophysical Research* 182, no. 4 (February 1):523–30.

Stuart, R. 1976. "Project is Moving National Power Grid Near Reality." *New York Times*, May 25.

Stys, Z.S. 1975. "Air Storage System Energy Transfer (ASSET)—Huntdorf Experience." In Proceedings of the Workshop on Compressed Air Energy Storage System, Sponsored by the Energy Research and Development Administration and the Electric Power Research Institute, Airlie House, Virginia, December 18–19.

Technology Review. 1978. "Retrieving Hydrogen Photochemically." (January): 26.

Telkes, M. 1975. "Thermal Storage for Solar Heating and Cooling." In Proceedings of the Workshop on Solar Energy Storage Subsystems for the Heating and Cooling of Buildings, NSF-RA-N-75-041, Charlottesville, Virginia, April 16-18, pp. 17-23.

Wentorf, R.H., Jr., and R.E. Hanneman. 1974. "Thermochemical Hydrogen Generation." *Science* 185 (July 26):311-19.

Wicklein, J. 1977. "Must We Try for Blackout III?" *The Progressive* (November):16-20.

Young, L.B. 1973. *Power Over People*. New York: Oxford University Press.

———. 1978. "Danger, High Voltage." *Environment* 20, no. 4 (May).

Young, L.B., and H.P. Young. 1974. "Pollution by Electrical Transmission." *Bulletin of the Atomic Scientists*, December, pp. 34-38.

Chapter 6

Energy Strategies

INTRODUCTION

Having assessed the future U.S. need for energy and having examined the key components of an overall energy supply system—including primary energy sources and conversion, storage, and transmission technologies—we are now able to address the task of integrating the various pieces into a plan for a workable energy future for the nation, free of many of the risks entailed in an attempt to sustain historic energy growth rates.

The present U.S. energy system, dependent upon nonrenewable fuels—oil, natural gas, and coal—for over 90 percent of its total supply rests, as we have seen, on a fragile foundation that is incapable of sustaining us very far into the future. Oil and gas, increasingly in short supply, together account for roughly three-fourths of the energy we use. It is inevitable that these fuels will become increasingly expensive in the future as the most accessible wells run dry, leaving only the more difficult (and hence more costly) accumulations to exploit. These too will eventually run dry, perhaps within a few decades. In the meantime, nearly half of the oil we consume is imported from foreign countries—a factor contributing in large part to presently imposing U.S. balance of trade deficit and creating a national security problem as well. Add to this the problems with nuclear power, now so grave that the prospects for this technology are at best dim. Couple all the difficulties with energy supplies with a lagging and, in part, misguided effort to develop alternative sources, a failure to use energy in frugal and efficient ways, and with an apparently insatiable national appetite for energy, and you have the principal ingredients of the U.S. energy dilemma.

How can the United States best resolve its massive energy problems? How can it shift from sources soon to be exhausted to sources not yet developed and do so in a timely way, with bearable social, political, and economic costs? A shift to coal does not offer a long-term solution to the nation's energy dilemma. Although U.S. reserves of coal are vast, capable of sustaining national energy use for one hundred or more years, the utilization of coal may ultimately have to be curtailed to prevent irreversible climatic modifications induced by cumulative additions of carbon dioxide into the atmosphere. Numerous other social, environmental, and climatic considerations also militate against an extensive reliance on coal. It has been estimated that if the annual rate of increase in the worldwide use of carbon-based fuels remains constant at the current level of 4.5 percent per year, the atmospheric carbon dioxide content would double roughly by 2035 (MacDonald 1979). More conservative estimates would place the date at which doubling would occur as sometime between 2050 and 2075. A doubling of atmospheric levels of carbon dioxide could lead to unacceptable global heating, increasing average global surface temperatures by about $5°F$ and causing significant climatic changes. Therefore, the goal of severely reducing, if not eliminating, the nation's dependence on fossil fuels roughly by the year 2050, with the hope that other major energy consuming nations will follow suit, appears to provide a critically important target for the coming energy transition.

For the long term, the United States can only choose among three major energy supply options—solar energy (including indirect forms such as wind, biomass, and hydropower), breeder reactors, and nuclear fusion. Only the first two methods are presently known to be within our technical capabilities. The technological feasibility of the third option—nuclear fusion—remains to be demonstrated. The simpler controlled fusion reactions currently under development, even if they work, will, as we have seen, have many undesirable features for commercial energy production. More advanced fusion reactions are considerably more difficult to harness, and it is thus impossible to guarantee their success. Therefore we can neither have confidence that fusion will work nor rule it out.

Since both breeder reactors and solar energy appear to be technically capable of supporting a high level of national demand far into the future, the choice between them should be based on economic, environmental, and social grounds. On the basis of presently available information, no clear-cut economic advantage can be claimed for either approach.

Breeder reactors, however, are an inherently inflexible technology capable only of producing electricity—a high quality and expensive energy form whose use is restricted by economic considerations to a limited set of appropriate applications. An energy supply system based largely on plutonium breeder reactors would also pose significant risks to the public owing to the potential for accidental releases of radioactivity and the deliberate diversion of fissile material for terrorist purposes at many stages of the nuclear fuel cycle. Furthermore, the vigilant administration of a large-scale plutonium economy may in all likelihood prove to be incompatible with the exercise of civil liberties and democratic control. Therefore, we consider a primary dependence on breeder reactors to be an unwise choice as the centerpiece of the nation's energy future.

On the other hand, we believe that most, if not all, of the major environmental and societal risks posed by a large-scale breeder future could be avoided in a well-planned solar energy economy. A solar energy future would in some ways be simpler and in other ways more complex than a breeder future. While the individual solar technologies are, in themselves, much less complicated than breeder reactors, the design and integration of a comprehensive solar energy system would involve a greater degree of sophistication than a system based entirely on breeders. However, the resulting system would be more flexible, less dangerous, better adjusted to the diverse energy needs of this complex industrial state, and thus a more stable basis for the nation's continued prosperity than any system based on nuclear fission.

It is worth noting that solar energy systems and breeder reactors are not mutually exclusive on technical grounds. An energy system based on a small number of breeder reactors operating in tandem with a broad range of solar technologies would almost surely present a greatly diminished threat as compared to a full-scale plutonium economy. However, it is not clear whether the nation could afford the costs of fully developing both solar and breeder technologies. Furthermore, the costs of research and development needed to produce a commercially viable breeder reactor not only promise to be high, but would also be largely independent of the ultimate size of the resultant industry. Therefore, the development of the breeder for a strictly supplemental role would entail an investment of highly questionable wisdom if it can be firmly established, as we believe it can, that renewable energy technologies are capable of performing the entire job more effectively and safely. In brief, there are most compelling practical reasons for judging that breeder-based and solar-based energy systems are mutually exclusive. It is clear that of the two, the solar option is much to be preferred if it can be shown to be both workable and affordable. In the following section, we outline a strategy for meeting the entire spectrum of future U.S. energy needs from a range of carefully selected solar technologies.

A SOLAR ENERGY FUTURE

A solar energy future implies diversity: no single solar technology can solve all our energy problems or satisfy all our needs. Thus, a broad mix of solar technologies—suited to both the natural energy income indigenous to a specific region and the required end use energy form—will be required. The choice of appropriate solar technologies must be mindful of the diverse scope of energy needs in an advanced industrial economy. Development of the necessary technologies must, furthermore, be practically realizable at an acceptable cost.

Owing to its high thermodynamic quality, solar energy is readily adaptable to a variety of end use tasks requiring various grades of energy. A combination of solar technologies can yield energy in all the requisite and useful forms—thermal and electrical energy and, with appropriate material inputs, liquid and gaseous fuels. However, physical constraints on the production of carbon-based liquid and gaseous fuels may in the long run force some shifts in consumption patterns with regard to the form of energy consumed, without necessarily affecting end uses.

In Chapter 2, we assessed future requirements for thermal energy—low temperature ($< 212°$F), intermediate temperature (212–$572°$F), and high temperature ($> 572°$F)—electricity, and liquid and gaseous fuels in the residential and commercial, industrial, and transportation sectors. It was found that with the adoption of the proposed energy efficiency improvements, energy use in the year 2050 could amount to no more than about 80 quads even while allowing for an increased population and a greatly elevated "energy standard of living" for all U.S. citizens. This estimated level of energy use is well below values normally arrived at in official projections and is only slightly above the 1978 level of energy use. We have concluded that with a shrewd selection of solar energy supply, storage, and transmission technologies, the full range of future energy requirements can be met. The core of our solar strategy is outlined in Tables 6–1 and 6–2 followed by a step-by-step discussion of how it would be possible to satisfy all our long-term energy needs by relying exclusively on renewable energy resources. Table 6–1 provides a breakdown of energy requirements for the various consuming sectors and matches appropriate renewable energy resources and technologies with the numerous energy-requiring tasks. Table 6–2 indicates possible energy supply and demand scenarios for the year 2050 assuming that the average level of energy efficiency in the United States is doubled by that year.

Thermal Energy

The future demand for relatively low temperature thermal energy to provide service hot water and space heating and cooling for residential and commercial buildings we estimate will comprise roughly 20 to 25 percent of total primary energy requirements in the long term. A mix of solar technologies will be needed to meet this energy requirement. A significant fraction of the energy needed for building space heating and cooling can be satisfied by passive solar techniques that employ energy-efficient construction and that utilize natural energy flows to move heat into and out of buildings as needed. Additional low temperature thermal energy for building and water heating (as well as for agricultural crop drying and a few industrial applications) can be provided by on-site solar energy systems of the "active" type (meaning that the mechanically forced circulation of a heat transfer fluid is relied on) employing simple flatplate or plastic solar collectors. Somewhat more sophisticated collectors, perhaps treated with selective surfaces or using "evacuated" tubes, are required to deliver the higher temperature heat needed to power solar air conditioners. The relatively small quantities of intermediate temperature (212–572°F) thermal energy needed for tasks such as cooking and clothes drying can be supplied by nontracking solar concentrators (such as the compound parabolic collector) or by burning a suitable solar-generated fuel such as methane or hydrogen.

The technology for passive and active solar heating and cooling is sufficiently well advanced that over 60,000 solar water heating, space heating, or cooling systems have been installed annually across the country in recent years. There clearly is no single "best" approach to providing low temperature heat to residential and commercial users; the optimal approach will vary from region to region and according to the specific need at hand. For example, in addition to solar energy systems designed to meet the needs of individual homes or buildings, community scale solar heating systems, perhaps utilizing large ponds for the collection and seasonal storage of solar energy, look quite promising. In many cases it might prove preferable to combine the

Table 6.1 A Proposed Long-term Solar Energy Economy.

Demand Sector	End Use Energy Form	Application	Percentage of Total Energy Use	Appropriate Energy Supply Technology
Residential and	Low Temperature Thermal Energy (<212°F)	Space Heating, Water Heating, Air Conditioning	20–25	Passive and Active Solar Systems, District Heating Systems
	Intermediate Temperature Thermal Energy (212–572°F)			Active Solar Heating with Concentrating Solar Collectors
	Hydrogen	Cooking and Drying	~5	Solar Thermal, Thermo-chemical, or Electrolytic Generation
	Methane			Biomass
Commercial	Electricity	Lighting, Appliances, Refrigeration	~10	Photovoltaic, Wind, Solar Thermal, Total Energy Systems
			Subtotal ~35	
	Intermediate Temperature Thermal Energy (<572°F)	Industrial and Agricultural Process Heat and Steam	~7.5	Active Solar Heating with Flatplate Collectors, and Tracking Solar Concentrators
	High Temperature (>572°F)	Industrial Process Heat and Steam	~17.5	Tracking, Concentrating Solar Collector Systems
Industrial	Hydrogen			Solar Thermal, Thermo-chemical, or Electrolytic Generation
	Electricity	Cogeneration, Electric, Drive, Electrolytic, and Electrochemical Processes	~10	Solar Thermal, Photovoltaic, Cogeneration, Wind systems

Feedstocks	Supply Carbon Sources to Chemical Industries		Biomass Residues and Wastes or Plantations
		~5	Biomass Residues and Wastes or Plantations
		Subtotal ~40	
Electricity	Electric Vehicles, Electric Rail	10-20	Photovoltaic, Wind and Solar, Thermal-Electric
Hydrogen	Aircraft Fuel, Land and Water Transportation Vehicles		Solar Thermal, Thermo-chemical, or Electrolytic Generation
Transportation	Long Distance Land and Water Transportation Vehicles	5-15	Biomass Residues and Wastes or Plantations
Liquid Fuels—Methanol, Ethanol, Gasoline			
		Subtotal ~ 25	
		100	

Source: UCS.

Table 6-2.   Energy Supply and Demand in the Year 2050.

End Use Energy Form	Energy Source	Appropriate Energy Supply Technology	Percentage	Energy Requirements (in quads) High Efficiency-High Population Scenarios		
				Current Standard of Living	Intermediate Standard of Living	High Standard of Living
Low Temperature Thermal Energy (<212°F)	Direct Solar Energy	Passive and Active Solar Heating and Cooling, District Heating Systems	25	13	17	20
Intermediate to High Temperature Thermal Energy (>212°F)		Flatplate Collectors, Stationary and Tracking Solar Concentrators	25	14	17	21
Electricity	Direct Solar Energy	Photovoltaic, Solar-Thermal, and Cogeneration Systems	30-40	8-10	10-13	12-20
	Wind	Wind Generators		8-11	10-14	12-20
		Subtotal		16-21	20-27	32
Liquid Fuels,	Biomass	Organic Residues and Wastes	10-20	3-5	3-7	3-5
Carbon Feedstocks, Methane		Energy "Plantations"		2-5	3-6	3-5
		Subtotal		5-10	6-13	8
		Total	100	53	67	81

separate tasks of providing heat and electricity by relying on "total energy" systems to supply the intermediate to low temperature heat left over from electrical generation from solar thermal-electric conversion devices, cogeneration units, or fuel cells. Again, district energy systems may offer a highly attractive approach for providing both the heating and the electricity needs of some communities.

Industrial and agricultural process heat and steam applications constitute a very significant component of end use energy requirements and may account for about 25 percent of total demand in the future. Stationary, and the more sophisticated tracking, solar collectors can deliver heat at the intermediate to high temperatures needed for these purposes. In most cases, thermal storage devices will be needed to insulate industrial production from fluctuations in the availability of sunlight. For those industrial applications where the direct use of solar heat is impractical, a suitable fuel such as solar-generated hydrogen or methane or, in some cases, electricity could be used.

Electrical Generation

Electricity will continue to be an important energy form in the long term and may account for about 30 to 40 percent of the gross energy consumed, depending on the degree to which a future transportation sector is electrified, as discussed below. Although an expanded role for electricity runs contrary to the predilections of some environmentalists, electrical generation represents an attractive method of harnessing the sun's energy that is both appropriate and efficient for many applications and that, furthermore, can be carried out without the risks or damage of most conventional electric-generating methods. Among the alternatives, wind, photovoltaic, and solar thermal-electric systems look the most promising and can produce electricity with minimal adverse side effects. A combination of these technologies should be capable of providing all long-term electricity requirements. Supplemental contributions can also be obtained from other solar-electric technologies, should additional supplies of electricity be needed.

The winds represent a vast source of energy that could be harnessed to provide up to 40 quads of primary energy annually should we need it. Wind machines are a proven technology and can provide the cheapest "solar" electricity today. In fact, wind systems are already nearly competitive with other sources of electricity in some regions of the country. In light of the fact that materials account for only a small fraction of the total cost of wind machines, a tremendous potential exists for reducing manufacturing costs (and simultaneously

expanding the market demand for wind systems) through the introduction of mass production techniques. Large investments of government capital could quickly bring the wind power industry to a state
of economic maturity. However, a few problems still need to be resolved. For example, barriers to linking wind generators with conventional utility grid systems need to be overcome. In addition, the
optimum size of wind machines has yet to be determined. Although
the federal wind program currently favors large machines, small systems are more suitable for mass production and consequently may
prove to be less costly. It appears likely that important roles will
exist in the future for both small and large wind systems.

A major contribution can ultimately be expected from photovoltaic or solar cells. These semiconductor devices directly convert
sunlight into electricity, offering unique advantages in terms of
clean, quiet, and trouble-free operation. Although cell prices are
presently high, major cost reductions have been achieved in related
semiconductor electronics industries, and many observers are confident that expanded production can lead to the availability of cost-
competitive cells within a decade. Large-scale government purchases
of solar cells can surely hasten the development of cost-effective
cells. There are a number of promising avenues to photovoltaic conversion involving a variety of possible materials and configurations,
and it is too early to tell which approaches will ultimately win out.
Nevertheless, the considerable progress achieved to date and the impressive rate at which novel ideas are being generated provide a
strong basis for optimism regarding the future role of photovoltaics.

Solar thermal conversion systems also offer important options
for electrical generation. However, these systems vary considerably
in conception, extending far beyond the relatively inflexible, large-
scale "power tower" design, which has received the most attention
and the majority of federal funding to date. Of enhanced appeal are
smaller systems, located close to their load, that can make use of
the residual "waste" heat from electrical generation. Community or
factory scale "total energy" systems providing thermal energy and
electricity, and storing the thermal energy on a seasonal basis, hold
considerable promise. Numerous configurations are possible. Heat
engines exploiting the temperature differentials in solar energy collection ponds may offer a simple and attractive solution for some
uses. More elaborate systems, relying on arrays of mirrors or on
tracking, parabolic solar collectors, may be appropriate in other
cases. The ultimate role assumed by solar thermal conversion systems will probably be defined largely on the basis of their relative
economic attractiveness and practicality as compared to photovoltaic systems.

The selection of an appropriate mix of wind, photovoltaic, or solar thermal-electric systems will depend on the relative abundance of wind energy and solar insolation within a specific region, as well as on the practicality and cost-effectiveness for the specific application at hand. Wind- and solar-electric systems should be viewed as complementary, rather than competitive, with substantial advantages—in terms of improved reliability and reduced storage requirements—to be secured through their interconnection.

Many other renewable energy technologies have been proposed for the generation of electricity, but our analysis indicates that none can confidently be counted on to supply substantial amounts of energy in the future.

Our scenarios do not anticipate appreciable contributions from hydroelectric power generation in the year 2050. If left undisturbed, dam reservoirs at the nation's major hydroelectric installations will gradually fill in due to continuous siltation, which will ultimately render them useless. Although consideration is being given to desilting these reservoirs, we have not assumed that these measures will be taken. Hydropower will, nonetheless, play an important role in the future as a means of providing utility load leveling and large-scale energy storage. However, despite the importance of these functions, neither contributes any net energy to the nation's supply.

The prospects for other proposed solar-electric generation schemes are even less favorable. For example, satellite power systems, which would beam solar electricity to earth in microwave form, appear unattractive on the basis of high development costs and numerous potential environmental hazards. Similarly, the technical and economic feasibility and the environmental acceptability of generating electricity by means of floating ocean thermal energy conversion (OTEC) plants remain to be verified. Tidal power systems, although technologically proven, are of extremely limited potential owing to the small number of suitable locations. Major technical and economic advances are needed before wave energy conversion can be considered a practicable option. Ocean currents do not appear to represent a power source that can be practically harnessed to produce significant amounts of energy. Finally, excessive demands for fresh water and other environmental problems will prohibit the extensive use of salinity gradients for power generation.

Transportation

Transportation in the postpetroleum era will pose a critical challenge for a solar energy economy. We expect the transportation sector to continue to consume roughly one-quarter of total long-term U.S. energy requirements. On the basis of a thorough review of the alternatives, we conclude that a mix of electricity, hydrogen, and alcohol fuels derived from biomass (organic matter) offers the most attractive approach to powering a future transportation sector.

Other suggested approaches look less satisfactory. For example, the use of liquid fuels derived from coal has been excluded from our scenarios for a number of reasons. First of all, coal is a nonrenewable fuel that, although abundant, is nonetheless incapable of sustaining consumption indefinitely into the future. In order to support present transportation energy requirements exclusively with liquified coal, overall coal production would have to be roughly trebled, leading to substantial environmental impacts. Furthermore, coal liquefaction is presently a highly expensive endeavor, and huge investments of capital would be required to establish a large-scale synthetic fuels industry.

An additional constraint is imposed by the anticipated climatic effect of continued, large-scale coal combustion. Therefore, coal-derived liquid fuels appear to be an unwise transportation option, except possibly as an interim measure on a moderate scale.

The use of a high temperature thermal energy storage device in conjunction with a heat engine such as a Stirling cycle also appears generally unfeasible owing to several drawbacks. The energy and power densities of thermal storage devices are generally quite low (Behrin et al, 1977). Highly efficient thermal insulation would be required to minimize energy losses over the extended periods of time when vehicles are not in use. Given that the average automobile is used only about one hour per day, the costs of providing adequate insulation would likely prove to be prohibitive. Recharging the thermal store would be an extremely slow process, causing an added inconvenience. Furthermore, because thermal energy can only be effectively transported over short distances, the distribution of energy in thermal form on the extensive scales needed would pose serious, if not insurmountable, difficulties.

Biomass-derived alcohols, such as methanol or ethanol, are attractive fuels for a future transportation sector, because they can almost directly substitute for liquid petroleum that presently provides the overwhelming majority (over 95 percent) of energy used in transportation. In fact, "gasohol"—a mixture consisting of about 90 percent gasoline and 10 ethyl alcohol (ethanol)—was being sold in more than 800 gas stations in the United States in 1979 (DOE 1979a). A much more ambitious program is currently underway in Brazil, where ethanol is expected to displace 7 percent of the nation's total gasoline consumption in 1979 and nearly 10 percent by 1980 (SEIR 1979a). The addition of ethanol to gasoline can clearly save petroleum and can also help boost the octane rating of regular gasoline. Using present technology, however, the energy required for the fermentation process and to distill the alcohol may exceed the energy content of the fuel produced (Chambers et al. 1978; McQuiston 1979). There is considerable disagreement over this point. A recent analysis indicates that ethanol can in fact be produced with a net energy gain (OTA 1979b). Methyl alcohol, or methanol, which can be produced more efficiently from less costly, nonfood sources, offers perhaps greater potential as a future transport fuel. However, fundamental constraints on biomass fuel production will probably limit these fuels to a supplementary, rather than dominant, role in transportation.

The principal limitation on "biofuel" production stems from the fact that green plants are relatively inefficient in converting sunlight into stored chemical energy by means of photosynthesis. A direct consequence is that unacceptably large amounts of land and irrigation water would have to be devoted to the cultivation of "energy crops" in order to run the entire transportation sector on alcohol fuels. The commitment of extensive amounts of land and water exclusively for this purpose would be indefensible in light of competing food requirements. The possibility of utilizing water-logged and presently unused lands for the cultivation of energy crops, however, needs to be investigated.

A more sensible approach than growing energy crops separately would be to integrate biomass fuel production with other activities—primarily agricultural and lumber production and solid waste disposal. For example, one proposal has been made to link the production of livestock with the generation of fuel alcohols and methane (Commoner 1979a). Under this arrangement, acreage currently devoted to growing grains and hay to support livestock production would instead be used to grow corn, sugar beets, and hay. Most of the corn and sugar beet crop would be fermented to produce alcohol, and the residue remaining after fermentation would be fed to livestock. Finally, biogas digesters would be used to produce methane from livestock manure. While this proposal is certainly intriguing, the extent to which the production of livestock feed grain would have to be expanded in order to produce both livestock feed and useful fuels simultaneously remains to be determined.

Liquid fuels could also be produced from organic wastes and residues from farms, logging industries, and urban refuse. While this resource should certainly be utilized to some extent—both to produce valuable fuels and simultaneously to help alleviate mounting environmental problems caused by the disposal of garbage, sewage, and sludge—it nonetheless suffers from inherent limitations. There are limitations on the net quantity of potentially available organic wastes, the fraction of these wastes that can practically be recovered, the percentage of agricultural and forestry wastes that can be utilized without impairing soil quality, and the efficiency at which organic matter can be converted into useful fuel forms such as methanol and methane. On the basis of very optimistic assumptions, the extractible energy potential in the form of liquid fuels from the U.S. biomass residue and waste resource is no more than 5 quads annually using conventional recovery and conversion methods. This amount of energy could support only a fraction of the total requirements of even a highly efficient transportation sector. Moreover, it is not yet clear whether this goal is practically achievable or whether it could be realized without significantly depleting the nutrient content of productive soils. A proposal to utilize solar heat in the conversion of organic matter into hydrogen, methanol, and even gasolinelike fuels (Antal 1976) may offer a promising means of extending the net yield of the biomass residue and waste resource. The ultimate practicability of this approach, however, needs to be verified, and thus any significant increase in the availability of biofuels from residues and wastes remains speculative.

In addition, a competing demand on the biomass resource would eventually be imposed by petrochemical industries in need of a dependable source of carbon feedstocks in the postpetroleum era. Although the magnitude of the petrochemical energy requirement can be reduced in the future, the wholesale elimination of this industry would appear unlikely, as well as imprudent. Given the fact that petrochemical industries specifically require a carbon source, while the transportation sector does not, it would make sense to allocate some portion of the renewable carbon feedstocks to appropriate chemical industries.

It is thus apparent that many unresolved questions obscure the ultimate role that biofuels could play in a future transportation sector. Uncertainties exist over whether appreciable quantities of fuels could be produced from biomass "plantations" without competing with other, more pressing demands on land and water use. The extent to which alcohol production could be combined with routine crop growing is also unclear, as is the net quantity of fuels that could be produced from organic residues and wastes. The picture is further clouded by the uncertain magnitude of the petrochemical feedstock requirement. Nonetheless, fuels from biomass will, at a minimum, be capable of making a small, but critical, contribution to our future transportation requirements. Priority allocations of liquid biofuels should be made to transportation vehicles for which no other practical fuel alternative exists. Included in this category would be agricultural vehicles and heavy trucks whose functions could not be readily substituted for by rail transport. If additional quantities of liquid fuels became available, they could sensibly be utilized for long distance automobile and bus travel and possibly, to a lesser extent, for water transport vehicles. In order to provide the remainder of the transportation requirements, we must turn to energy forms whose supplies are subject to less rigid constraints than fuel alcohols.

The combined use of electricity and hydrogen appears capable of meeting all remaining energy demands in transportation, beyond that provided by biomass sources. Energy in these forms can be made available in quantities sufficient for meeting demands in the transportation sector, as well as for other appropriate end use applications. Electricity can be generated on a sustainable basis from a number of solar- and wind-electric technologies. Hydrogen can be produced from solar electricity by means of the electrolysis of water, or it may be generated by solar thermal techniques involving either the thermal splitting of water or thermochemical reactions. Both of these approaches utilize solar energy with much greater efficiency—roughly by factors of ten to a hundred—than the biomass plantation approach.[1]

Electricity is well suited for powering rail transportation and urban mass transit systems. Battery-powered electric vehicles should become attractive in the long term, especially for short distance travel, which accounts for the overwhelming majority of automobile trips. In fact, it has been estimated that roughly 90 percent of all automobile trips by U.S. drivers are less than twenty miles (Kihss 1979) and more than 50 percent of all trips are for distances of five miles or less (Bossong 1978). Electric vehicles are less expensive to operate than vehicles powered by gasoline engines, because fuel and maintenance costs are much lower. Therefore, electric vehicles may be economical on a "life cycle" basis, even when initial costs are

---

1. For example, electricity—an energy form suitable for motive applications that is readily converted into mechanical power—can now be generated using silicon solar cells with an efficiency of 15 percent. The efficiency of hydrogen use—assuming it is thermochemically generated with a 25 percent efficiency and subsequently converted to electricity in 60 percent efficient fuel cells—would also be roughly 15 percent. Biomass energy can be photosynthetically derived from sunlight with an efficiency of perhaps 1 to 3 percent under highly favorable conditions. If the organic matter is then converted to methanol with a 50 percent efficiency and subsequently converted to driveshaft power in a heat engine with an optimistic 40 percent efficiency, the overall efficiency of the biomass plantation method would range from about 0.2 to 0.6 percent, a factor of twenty-tive to seventy-five below the efficiency values estimated for electricity and hydrogen.

twice those of conventional vehicles (NAS 1976; Bossong 1978). Another potential benefit stems from the fact that the overall costs of providing electricity could be reduced if the vehicles' batteries are recharged during off-peak hours.

Thousands of electric vehicles—primarily industrial vans and fork-lifts—are presently in use in the United States. In addition, the U.S. Postal Service presently has a fleet of several hundred electric vans and has recently decided to triple the size of the fleet to more than 1,100 vehicles (*New York Times* 1979). Several U.S. companies are planning for the assembly line production of electric automobiles in the mid-1980s and it has been predicted that significant numbers of electric vehicles—designed for special purposes and limited use— may be on the road at that time (OTA 1979). While additional work is needed to produce economical batteries with higher energy and power densities capable of powering high performance electric vehicles, considerable progress has been achieved to date, and a number of advanced primary and secondary batteries now look promising for this application. Japan, for example, announced in 1977 the development of an electric car that could achieve a speed of 50 miles per hour and travel a distance of 250 miles on a single battery charge (Bossong 1978).

Liquid hydrogen appears to be the fuel of choice for aircraft. It may also be used for water travel, although another liquid fuel may be more desirable when long shipping distances and times are involved, in order to avoid the need for elaborate insulation required to store the liquid hydrogen at the extremely low temperatures required. Hydrogen may also become an attractive fuel for use in some land vehicles, but problems involved with fuel handling and the on-board storage of hydrogen need to be overcome. Owing to the typically long periods of inactivity for most vehicles, liquid hydrogen—which must be maintained at a very low temperature— would be an unsuitable fuel form. Safety considerations involved in the dispensation and on-board storage of liquid hydrogen would also militate against its extensive use in automobiles. Metal hydrides offer a more practical method of storing hydrogen for most vehicular applications but are quite heavy. While metal hydrides cannot store as much energy per unit weight as liquid hydrogen systems, they do not appear to pose any significant safety hazards. One U.S. company has installed hydride tanks in several modified automobiles and buses to enable them to successfully run on hydrogen fuel (SED 1977a). A high pressure storage tank has been developed in Germany that can store enough hydrogen gas to enable a Volkswagen to travel a distance of 200 miles (SED 1977b). Most of the hydrogen-powered vehicles built so far have employed modified internal combustion engines. In the future, more efficient fuel cells may be used to convert hydrogen to electricity to power transportation vehicles.

The combined transportation role of electricity and hydrogen will be defined largely on the basis of the availability of liquid fuels derived from biomass. It is still too early to predict the optimal balance between electricity and hydrogen in a future transportation sector. Electricity clearly represents the most logical choice for rail transportation, urban mass transit, and short-range, inner city vehicles. Hydrogen appears to be an attractive fuel for aircraft. For the remaining transportation applications, the future mix of electricity and hydrogen will be determined on grounds of relative costs, efficiencies, practicality, and safety of the two approaches— all of which are, to some extent, uncertain at the present time.

Energy Storage

Because of the intermittent nature of solar and wind sources, energy storage will be an essential feature of a renewable energy economy. A range of storage techniques will be required to balance variations in output from solar and wind energy-gathering technologies against a variable energy demand. Energy will need to be stored in various forms that are compatible both with the energy source and the end use. Storage will be required over a variety of scales and time frames. The numerous proposed storage technologies have been reviewed extensively in Chapter 5.

A variety of approaches will ultimately be needed to store thermal energy, representing roughly one-half of future energy demand. Thermal energy may also be temporarily stored and transported in chemical form. A number of options are also available for storing electricity and mechanical energy. Large-scale storage can be accomplished using pumped hydroelectric and compressed air systems. Batteries and flywheels offer an important means of storage for both utility and transportation applications. Superconducting magnets may ultimately provide an attractive means of storing electricity directly. Solar-derived fuels can be stored relatively simply in comparison to other forms of solar energy. Gaseous fuels, such as methane and hydrogen, may be stored in underground reservoirs such as depleted gas wells and, to a limited extent, in pipeline distribution systems. Liquid fuel alcohols can be conveniently stored, transported, and dispensed in a manner similar to petroleum.

Energy Transmission

Energy transmission, needed to connect optimally sited solar energy collection and conversion facilities with energy use centers, is another essential component of a workable national energy system that is curiously neglected in most other proposed solar strategies. Some people cling to a vision of a "pastoral" energy future consisting almost exclusively of "decentralized," generation facilities at the point of end use in which transmission would be superfluous. While the on-site utilization of solar energy is generally preferable when feasible, some degree of transmission will be absolutely necessary to provide energy for concentrated urban and industrial centers where the density of energy use exceeds the local availability of solar and wind energy. For example, the density of power use in Manhattan exceeds the total incident flux of solar energy by a factor of roughly three and a half (SMIC 1971). Thus, even if all of Manhattan were covered with solar collectors, assuming present consumption patterns, the major fraction of the energy used would still have to come from outlying regions.

Apart from being simply a necessity, an intelligently planned energy transmission system can yield substantial benefits as well. For example, the interconnection of geographically dispersed solar and wind electric-generating facilities could have a profound effect on minimizing electrical storage requirements. Detailed studies are needed to weigh the advantages of siting generation facilities at locations having high average solar insolation or wind speeds against the incremental cost of transmission to consuming centers. The economic and environmental trade-offs between expanding storage capacity and an increased reliance on energy transmission also need to be explored.

A number of options for transporting energy over a range of distances and in various forms—thermal energy, electricity, and gaseous, liquid, and perhaps chemical fuels—will prove useful in the long term. Thermal energy can be economically transported in insulated pipes over distances of up to twenty-five miles in district heating and total energy systems (Karkheck et al. 1977). Given our expectation of an expanded role for electricity in transportation, electrical transmission should continue to be important far into the future. While the construction of a nationwide electricity grid system is already well advanced, further improvements can be made. In the short run, the bolstering of weak links in the existing grid system could both increase the reliability of electric supply and also alleviate, to come extent, the need to install additional peaking power generation facilities. Looking farther into the future, the prospect of a fully integrated national grid system—linking power pools of varying scales, each composed of an interconnected network of solar and wind generators—appears highly attractive. Ideally, the long distance transmission of electricity would take place in underground—and possibly superconducting—cables.

A fairly extensive natural gas pipeline distribution system is already in place and could be used without modification to transport solar-generated methane. Experience gained in Europe, and to a lesser extent in the United States, demonstrates that pure hydrogen gas can also be readily transported over long distances by pipeline. If hydrogen is extensively used, its optimal transmission would ultimately require the construction of new pipelines, apart from the existing natural gas pipeline system. In addition, biomass-derived liquid alcohols could be transported from areas of production to centers of use by means of pipelines and rail and truck tankers in a manner paralleling the present distribution of liquid petroleum. Finally, chemical "heat" pipes currently under study might eventually be used for the long distance transmission of solar thermal energy, temporarily stored as chemical energy in the form of solar-generated chemical reaction products (C&EN 1979a).

These, then, are the essential components of a long-range solar economy—technologies needed to harness the natural flow of solar energy in various forms, technologies needed to store energy in order to match fluctuations in energy supply against a variable demand, and facilities needed to transport energy from the point of collection to the site of use. We conclude that intelligently selected solar technologies can be integrated into a workable energy system for the United States.

Having acquired some perspective as to the shape and structure of a long-term solar economy, let us next consider the transitional period during which solar energy can gradually come into general use. The near-term contribution that might be expected from solar energy will be examined, and programs needed to guide national energy policy from the present to the long-range goal will be explored.

THE SOLAR TRANSITION

We believe that the United States can readily achieve a transition to a fully solar-powered economy by the year 2050. The precise duration of the transitional period is, of course, flexible rather than rigidly fixed and could be lengthened or shortened as required.

Nonetheless, the seventy-year period we have allotted for the solar transition appears reasonable when viewed in the light of the previous energy transitions experienced in this country within the past 130 years. For example, from 1850 to 1900, coal displaced wood as the dominant energy source in the United States, increasing its share of the nation's energy supply from about 9 to 72 percent. Oil similarly replaced coal as the nation's major energy source by 1950, its overall contribution increasing from about 6 to 40 percent since 1910.

Clearly, there are limits to the speed with which the transformation to a solar economy can occur. Solar technologies should be introduced with care, skillfully integrated with conventional energy and electricity delivery systems. The numerous technologies that will ultimately be needed should be phased in at intervals in keeping with their varying states of commercial readiness. In many cases, growth would occur slowly at first. As the production of solar devices gradually increased, resultant reductions in manufacturing costs would expand their potential market, possibly triggering a precipitous growth in use.

The most immediate payback can be realized by taking advantage of the numerous available and inexpensive means of increasing the efficiency of energy use. Important savings of energy can be made by retrofitting existing residential and commercial building for energy conservation. The energy requirements of new buildings could be considerably reduced by design improvements, but energy savings in this area would grow slowly because building turnover times tend to be long—often over fifty years. Reductions in energy use can be achieved more quickly in the industrial sector, where equipment turnover times are typically much shorter than for residential and commercial buildings. Significant amounts of energy can also be saved in many cases by modifying existing manufacturing equipment and techniques, rather than replacing them outright. The transportation sector offers the potential for the most rapid improvements in energy efficiency, owing to the generally short useful lifetimes of transportation vehicles, particularly automobiles, which are often less than ten years.

Solar technologies capable of providing low temperature heat for buildings, water heating, and agricultural and industrial applications are already well developed and will be among the first to be introduced on a large scale. Active and passive solar heating systems have already been installed in thousands of homes and buildings. These systems are presently cost-competitive with electric heating in many parts of the country and are thus quite ready for expanded commercial use. Solar air-conditioning systems are also commercially available, but it may take several years before they become generally attractive on an economic basis. Solar heating systems introduced at an early time might be installed with only limited storage capacity, relying on conventional fuels such as oil and gas or utility-supplied electricity to provide backup energy when the amount of solar energy collected is insufficient to meet demand. Ultimately, completely independent solar energy systems can be developed that economically store excess thermal energy on an annual basis—perhaps in underground formations such as aquifers or in covered and insulated surface ponds—or rely on solar-generated electricity or fuels such as hydrogen or methane as an auxiliary energy supply. Most of the systems installed initially will continue to be designed for individual buildings. The development of community scale solar heating systems with centralized thermal storage may offer an attractive alternative for future installations.

While the use of solar energy to provide low and intermediate temperature agricultural and industrial process heat appears to be an attractive application, the contribution made in this area so far has been quite limited, as the benefits of solar utilization need to be more clearly demonstrated to prospective investors. Higher temperature solar thermal systems will require somewhat more developmental work prior to their widespread introduction, but can almost certainly be made available within a decade.

While considerable progress is needed before many solar technologies can come into widespread use, there are notable exceptions. For example, small but important enlargements in hydroelectric power generation—by far the dominant renewable source of electricity today—can be made economically by taking advantage of presently unused capacity at existing small dams throughout the country. The other major role for hydroelectric power will be as a large-scale means of storing energy in both aboveground and underground facilities. Wind generating systems are technically feasible and beginning to look economically attractive at favorable sites in the United States. Although wind machines are only slowly being installed at various locations across the country, the commitment of generous amounts of government funds could quickly make a major wind power industry a commercial reality. While photovoltaic or solar cells are presently too expensive for all but a few limited uses, the market for these devices should expand considerably as a result of cost reductions expected by the mid- to late 1980s. A potentially explosive growth in sales could then ensue. Solar thermal conversion systems should also assume an important role in the future, but the optimal approach remains to be identified. While community scale solar power systems providing both heat and electricity look promising, development is only at a preliminary stage. No major technical breakthroughs are required, however. Many other proposed solar-electric technologies face major technical, economical, or environmental hurdles, and thus concete timetables for their introduction cannot be confidently set at present.

Changes in the present transportation sector will come gradually, unless supplies of petroleum—which almost alone powers all transportation in the United States—suddenly and unexpectedly dry up. Alcohol from agricultural biomass sources is already being produced in minute quantities and used to extend modestly the supply of gasoline. Production of biomass-derived fuels will continue to expand over time until constrained by fundamental limitations imposed on photosynthetic energy conversion. Electric vehicles are used today only for a very limited number of applications. Their widespread production awaits the development of advanced electric batteries capable of approaching the performance of gasoline-powered engines—not expected until 1985 at the earliest. Thus, the use of general purpose electric vehicles will probably not begin to grow appreciably before 1990. In the meantime, hybrid vehicles—powered by a small gasoline engine and electric batteries—may be relied upon to reduce oil consumption. Major improvements in the rail transportation system will be required before trains can begin to displace trucks for some fraction of freight transport and jets and airplanes for passenger travel. The use of hydrogen as a jet fuel will probably not occur until petroleum-derived fuels become prohibitively expensive and difficult to acquire. In any event, the widespread use of hydrogen for transportation could not occur until facilities for large-scale hydrogen production, storage, and distribution are established—an eventuality that is unlikely in this century.

The integration of solar technologies with conventional energy supply systems poses a major technical challenge for solar energy utilization. Until wholly autonomous solar energy systems are developed, an adroit blending and meshing of solar and conventional energy technologies will be required. Initially, solar energy systems may be built with only limited storage capability, relying on conventional fuels to provide energy when available solar energy is insufficient to meet demands. For example, solar heating systems could rely on oil or gas heating for backup. The operation of these systems would save fuel, but not necessarily reduce requirements placed on conventional systems. Similarly, electricity produced intermittently from solar- and wind-electric generators operating without storage could be fed into a conventional utility grid system. In fact, devices called synchronous inverters are already available that can convert solar- and wind-generated electricity into forms and frequencies compatible with our existing transmission and distribution system.

The fact that many storage systems are not commercially available today should not pose any near-term constaints to solar energy utilization. It has been estimated that solar- and wind-electric systems operating without storage, and hence variable in output, could feed their electricity into a conventional utility grid system without undermining overall grid stability until they contributed more than 10 to 20 percent of total electricity production. Thus, as the need for storage will grow gradually, in phase with the increased utilization of solar energy, more than sufficient time will be available for the development of cost-effective storage technologies.

The numerous storage technologies that will ultimately be needed are presently at varying stages of development. Many of the relatively simple approaches to thermal storage are technically and economically feasible today, but await the arrival of applications for which they are needed. On the other hand, techniques for storing electricity directly or indirectly are, by and large, less well advanced. A notable exception, however, is pumped hydroelectric storage, which is already used extensively in the U.S. utility systems. Pumped-hydro facilities could provide much of the storage capability needed in the short run and could be particularly helpful in spurring the development of wind energy systems, owing to the fact that many major hydroelectric installations are located in regions that are rather windy. In fact, a large windmill planned to be built at Medicine Bow, Wyoming, for the Bureau of Reclamation will be tied into the regional hydroelectricity grid; operation is expected to begin in 1981. Technology for compressed gas storage is also reasonably well established, with one large-scale system already completed in West Germany. Furthermore, considerable progress has been made in the development of advanced storage batteries, and there is a high probability that these devices will be ready for commercialization by the mid-1980s.

As the dependability of energy delivery from solar energy systems is gradually increased through enlargements of storage capacity, these systems will begin to displace conventional energy and electricity supply capacity. The eventual development of economical methods of storing energy over extended periods could lead to the emergence of completely independent solar energy systems. Auxiliary energy, when needed, could be provided from a number of renewable sources. For example, solar-powered electric heat pumps could provide supplemental heat for buildings. Similarly, solar-derived gaseous fuels such as methane or hydrogen could be transported by pipeline and converted into electricity and useful heat by means of cogeneration devices or fuel cells.

The most important "bridging" fuels to carry us through the solar transition will be those that we are currently most dependent on—oil and natural gas by virtue of their relative convenience, and coal by virtue of its abundance. Solar thermal technology can gradually supplant the use of fossil fuels for providing building and water heating and industrial process heat and steam. Similarly, solar-electric generation technologies can eventually substitute for fossil fuel power plants, which currently supply about 75 percent of our electricity. Some of the oil now used for transportation can eventually be displaced by biomass-derived alcohol fuels, but inherent limitations on the biomass resource will force a reliance on other transportation approaches, as discussed previously. Methane, the major constituent of natural gas, can be produced by the anaerobic decomposition of organic matter. It can readily substitute for, or be mixed with, natural gas using the existing pipeline distribution system. The relative proportion of solar methane to natural gas flowing in the pipeline would gradually increase over time. However, the amount of methane that could be produced from organic sources is limited by fundamental constraints on photosynthetic energy production similarly affecting the availability of liquid biofuels. Supplies of solar methane could be augmented by the addition of solar-produced hydrogen up to a point at which hydrogen comprised 10 to 20 percent of the overall mixture without requiring significant modifications in equipment designed for natural gas storage, distribution, or utilization (Commoner 1979b). If hydrogen were ultimately used more extensively to overcome limitations imposed on methane production, new pipelines for its transport would have to be built, and new conversion devices such as fuel cells would be needed for its utilization.

The use of oil, gas, and coal during the transitional period should proceed in a manner designed to maximize energy efficiency and minimize environmental disruption. Cogeneration or total energy systems offer attractive means of utilization in cases where both heat and electricity are required. Solar methane could readily fuel cogeneration units originally operating on natural gas. Solidified urban trash might be burned in fluidized bed combustion systems designed for the clean burning of coal. Fuel cells offer another promising approach to direct energy conversion. These devices can effectively operate on oil, gas, and coal-derived fuels, and ultimately, with even greater efficiency, on solar-produced hydrogen. Waste heat from the cells can be utilized, leading to very high overall efficiencies. With careful planning, a smooth transition can be made from oil, gas, and coal—the fuels that presently provide over 90 percent of our energy—to a wholly solar-powered economy.

POTENTIAL CONTRIBUTION OF SOLAR ENERGY IN THE YEAR 2000

In 1978, the United States derived roughly 4.8 quads, or about 6 percent of its total energy, from solar energy in the form of hydro-electricity and wood. As noted previously, we believe it is entirely plausible to assume that a transition to a solar economy could be completed by 2050. It is also important, however, to assess the contribution that could be expected from solar energy in the near term in order to judiciously plan for the interim use of conventional energy supplies. In this section, we review the estimates that have been made for a variety of solar technologies to provide a range of possible contributions for the year 2000.

Numerous estimates have been made as to the amount of energy that could be provided in the year 2000 by residential and commercial solar heating and cooling systems, and we find a range of about 2 to 4 quads to be quite reasonable (CEQ 1978; DOE 1978, MITRE 1978; SRI 1978). It has been estimated that by the turn of the century there will be more than one hundred million dwellings in the United States, nearly half of which will have been constructed between now and then (DOE 1978). Assuming that one-half of the newly constructed residential units are equipped with passive and active solar devices and that an additional one-third of the existing residential building stock are retrofitted for solar space heating and cooling, the net contribution from solar energy would amount to about 2 quads. A more aggressive implementation program would, of course, lead to a greater solar contribution. Additional amounts beyond 2 quads annually could be provided by the use of solar heating and cooling systems in commercial buildings.

The potential penetration of solar technologies into industry has been less fully studied. One comprehensive study prepared for the Department of Energy concluded that solar energy could provide 7.5 quads for intermediate temperature industrial process heat applications by the year 2000 (ITC 1977). Other estimates range from about 2 to 5 quads (CEQ 1978; DOE 1978; MITRE 1978). We have more conservatively assumed a range of 1 to 4 quads in our scenarios for the year 2000.

Hydroelectric power generation—by far the most important current means of renewable electricity production—cannot be significantly expanded in this country. At most, hydroelectric generation could probably be increased by about 1 quad over the present average value of 3 quads per year (DOE 1978). This expansion could be most economically achieved, with a minimal environmental disruption, by upgrading the generating capability at thousands of existing small dam sites throughout the country and be installing additional small-scale hydro systems if necessary. Thus, we conclude that 3 to 4 quads can be obtained from hydropower generation in the year 2000.

Wind power could make a much more significant contribution to the nation's electricity supply in this century than other new sources. Estimated contributions for the year 2000 range anywhere from about 2 to 20 quads (CEQ 1978; Coty and Dubey 1976; DOE 1978; FEA 1974a; Heronemus 1973; MITRE 1978; NSF/NASA 1972). While the latter figure appears to be unrealistically high, we conclude that a range of 2 to 5 quads is readily attainable.

According to Inglis (1978), for example, mass-produced 40 kWe wind machines are roughly comparable to a small automobile in terms of cost, weight, and complexity. Given that ten million cars are produced in Detroit every year, the annual production of wind machines on a scale averaging only 5 percent of the automobile industry in Detroit could, over a twenty-year period beginning in 1980, lead to an installed wind generation capacity of 400 million kWe in the year 2000. Assuming that these wind turbines were operated with a 25 percent average capacity factor, approximately 9 quads of primary energy could be provided annually.

While an implementation program of this magnitude would be admittedly ambitious, it is not without historical precedent. For example, over an average fifteen year period, U.S. electric utility companies typically install approximately 300,000 transmission towers, each of which is comparable in size to a tower that could support a 1 MWe wind turbine (DOE 1978). Assuming that 1 MWe wind machines were deployed at this same rate starting in 1980, the installed wind generating capacity would attain a level of 400 million kWe by the year 2000. Therefore, the logistical problems involved in the manufacture and installation of this many wind machines should not prove to be unmanageable.

The near-term contribution from photovoltaics will be largely dictated by the rate at which manufacturing cost reductions can be achieved. Predictions of the energy that could be supplied by solar cells in the year 2000 range from 1 to 8 quads (CEQ 1978; DOE 1978; ERDA 1975; FEA 1974a). Given the additional progress still required to produce economic cells, the latter figure may be overly optimistic. Nonetheless, with adequate economic stimulation provided, perhaps by means of an assured government market, an industrial photovoltaic production capacity could be scaled up quickly, making a sizeable contribution of several quads by the turn of the century entirely plausible. Our scenarios anticipate that photovoltaic systems could provide 1 to 4 quads in the year 2000.

The prospects for solar thermal conversion systems are more difficult to predict. Estimates made of the possible contribution that could be made by solar power plants in the year 2000 vary from 0 to 2 quads (CEQ 1978; DOE 1978; FEA 1974a; MITRE 1978), a range we find reasonable. Approximately 450, 100 MWe plants, large by solar power standards, would be needed to provide 1 quad of primary energy. Assuming that construction began in 1981, with a typical construction period of 10 years, an average of forty-five plants of this size would have to be built annually in order to achieve this goal by the year 2000. Of course, it is not unlikely that systems rated at only a small fraction of 100 megawatts may prove preferable for the majority of cases. If smaller size plants were indeed adopted, a correspondingly larger number would be needed.

Although many other solar-electric technologies have been proposed—including satellite power stations, ocean thermal conversion plants, and wave and tidal power generators—we conclude that the prospects of obtaining appreciable contributions from these sources in this century are very remote.

In 1978, the United States obtained roughly 1.8 quads of energy from biomass, primarily used in the form of residues from the wood products industries or as firewood. Of the numerous estimates made for the maximum contribution from biomass in the year 2000, the most reasonable lie within a range of 3 to 5 quads (CEQ 1978; CONAES 1978; DOE 1978).

Summing up our estimates for the various solar technologies yields a range of about 12 to 28 quads for the maximum potential contribution from solar energy in the year 2000, consistent with numerous other estimates that have been made (CEQ 1978; DOE 1979b; ERDA 1975; FEA 1974a; MITRE 1978; Rodberg 1979; SRI 1977, 1978). Our estimates are summarized in Table 6–3.

Table 6-3.   Potential Contribution of Solar Energy in the Year 2000.

		Contributions in Quads
Solar Thermal:	Residential and Commercial Heating and Cooling	2-4
	Industrial	1-4
Electric:	Hydroelectric	3-4
	Wind	2-5
	Photovoltaic	1-4
	Thermal-Electric	0-2
Biomass:	Residues and Wastes; "Plantations"	3-5
	Total	12-28

ENERGY SUPPLY IN THE YEAR 2000

In Chapter 2 it was found that with the adoption of our recommended efficiency improvements, energy demand in the year 2000 would amount to no more than about 100 quads, even though effective per capita energy use increased by 55 percent—allowing all income groups to enjoy the energy use levels now characteristic of only the upper 20 percent income group in the country. If effective per capita energy use were to remain unchanged on the average, enhanced energy efficiency would actually reduce the total amount of primary energy required in the year 2000 by about 15 percent as compared to 1978. On the basis of our estimates for solar implementation, solar energy would contribute roughly 12 to 28 percent of the overall energy supply in the year 2000, assuming that demand reached a maximum level of about 100 quads, and 15 to 35 percent if demand were held to an "intermediate" level of about 80 quads. This is consistent with estimates made that solar energy could provide 12 percent of the energy needed by 2000 under existing programs without a redirection of national policies (C&EN 1977b; Rattner 1979) and a maximum of 20 to 25 percent of the total supply, according to reports by the Department of Energy (DOE 1979b) and the Council on Environmental Quality (CEQ 1978).

While an admittedly ambitious program would be needed to expand the net contribution of solar energy from 6 percent of the national supply in 1978 to about 25 percent in the year 2000, an increase of this magnitude is not wholly unprecedented. For example, the use of oil grew from providing 6 percent of national energy use in 1910 up to 23 percent in 1930. Similarly, the net contribution of natural gas increased by about 18 percent overall, from about 11 percent of the total supply in 1940 up to about 29 percent in 1960 (Department of the Interior 1975).

In addition to solar energy, another "nonconventional" resource, geothermal energy, could contribute small amounts of energy to the nation's supply in the year 2000. However, major uncertainties exist in assessing this resource. Estimates of installed geothermal electric-generating capacity in the year 2000 extend upwards as high as 100,000 to 400,000 MWE (ERDA 1975; FEA 1974b). Assuming a capacity factor of 60 percent, this installed generating capacity could annually produce about 5 to 20 quads of primary energy. However, generation at this rate could exhaust identified resources of accessible, high temperature geothermal liquids and vapors within a few decades. Therefore, we find a net annual contribution from geothermal resources of about 2 quads by the year 2000 to be a more cautious estimate.

The implications of solar energy providing a significant fraction of our energy in the year 2000 are truly striking when viewed in terms of alleviating demands placed on conventional, nonrenewable energy resources. Various energy supply and demand scenarios for the year 2000 are given in Table 6–4. All scenarios assume a 25 percent improvement in average energy efficiency and a high rate of population growth through the year 2000. It is important to note that in all but one scenario, total demand for fossil fuels—coal, oil, and natural gas—is reduced from 1978 levels by anywhere from about 6 percent to more than 50 percent. The single exception is the case in which energy demand reaches a maximum level of 102 quads, owing to 55 percent average increases in effective per capita energy use. If a low rate of solar implementation is assumed, demand for fossil fuels in this scenario will increase by about 14 quads or 19 percent. Assuming oil and gas production could be maintained at current levels, this increase could be achieved precisely by doubling coal production—a prospect anticipated by 1990 in Department of Energy (DOE 1979c) forecasts. However, if a high rate of solar utilization were achieved, aggregate fossil fuel demand in this high energy use scenario would actually decrease by about 7 percent. It must also be emphasized that overall decreases in fossil fuel consumption could be achieved without any increases in nuclear energy production. Total energy demand in the year 2000 could be met with the contribution from nuclear power either held at the 1978 level or, in some cases, reduced to zero. If the 190-odd nuclear plants either currently licensed, under construction, or on

Table 6-4. Energy Supply and Demand in the Year 2000.[a]

Energy Source	1978	2000 "Current" Standard of Living[b]		2000 "Intermediate" Standard of Living[b]		2000 "High" Standard of Living[b]	
		"Low" Solar	"High" Solar	"Low" Solar	"High" Solar	"Low" Solar	"High" Solar
Nonrenewable							
Fossil, Coal, Oil, and Natural Gas	71.8	52	36	67	54	85	69
Nuclear (LWR)	3.0	—	—	3	—	3	3
Geothermal	0.07	2	2	2	2	2	2
Renewable							
Solar Thermal, Building Heating and Cooling		2	4	2	4	2	4
Industrial		1	4	1	4	1	4
Solar Electric:							
Hydro	3.1	3	4	3	4	3	4
Wind		2	5	2	5	2	5
Photovoltaic		1	4	1	4	1	4
Thermal-Electric		—	2	—	2	—	2
Biomass	1.8	3	5	3	5	3	5
Total	79.8	66	66	84	84	102	102

[a]All scenarios assume the adoption of energy efficiency measures designed to reduce the amount of energy required to perform a given task by an average of 25 percent. A "high" population growth rate is also assumed, corresponding to a 23 percent net increase by the year 2000.

[b]The "current" standard of living case assumes no overall change in effective per capita energy use, while the "intermediate" and "high" cases assume increases of 28 percent and 55 percent, respectively.

order were in fact built and operated, primary energy produced from
nuclear generation would increase from the current annual level of
3 quads up to about 10 quads, correspondingly reducing requirements
for fossil fuel power generation.

Without major increases in effective per capita energy use, energy
demand in the year 2000 could be as low as about 60 quads, assuming the efficiency of energy utilization were increased by 25 percent.
With solar energy providing 15 quads, or 25 percent of total energy
use, only 45 quads of nonrenewable energy would be required—a
40 percent drop from current levels. Such a low level of demand
would allow for considerable flexibility in fulfilling our energy

Figure 6-1.   Transitional Path to a Solar Economy.

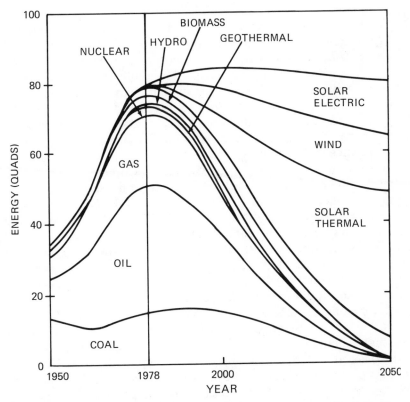

(High Solar Implementation Scenario; High Efficiency; High Population; Intermediate Standard of Living by 2000; High Standard of living by 2050.)

requirements. If necessary or desirable, energy production from all nonrenewable sources could be reduced from current levels.

Beyond the year 2000, with the introduction of solar technologies accelerating, the contribution from solar energy would become dominant, greatly reducing pressure on all nonrenewable energy sources—particularly oil and gas, which would be those in shortest supply. The target date we have set for an all-solar economy is the year 2050. A possible transition path is depicted in Figure 6-1. This scenario assumes the following: a rapid rate of solar deployment, overall energy efficiency improvements of 25 percent by the year 2000 and 50 percent by 2050, a high rate of population growth, and standard of living increases raising effective average per capita energy use by 28 percent by the year 2000 and 55 percent by the year 2050.

INSTITUTIONAL STRATEGIES

The present system of providing generous subsidies to conventional energy sources has distorted market prices and obscures the potential economic advantages offered by energy conservation and solar energy utilization. Many institutional roadblocks—pricing policies, tax laws, federal research and development priorities, building codes, zoning ordinances, and so forth—inhibit the adoption of conservation measures and the installation of solar devices. A reorientation of current policies is needed to promote the efficient use of energy and a reliance on solar energy.

Several options are available. The strategy that has received the greatest emphasis by the Carter administration has been to rely on taxes and increased prices for oil and gas to induce "conservation." This strategy is aimed at correcting the artificially low prices brought about by regulated controls on the price of oil, gas, and electricity and by the practice of averaged pricing, whereby the price of producing new oil or electricity from newly constructed plants, for example, is averaged with the cheaper price of old oil or electricity from already built units. This practice disguises the replacement costs for conventional energy supplies and provides a strong incentive for their expansion even though the incremental costs of expansion may be uneconomic when compared to the costs of conservation and solar utilization. Therefore, a pricing policy that accurately reflects energy replacement costs is widely believed to be an essential prerequisite to a reliance on market forces to encourage energy efficiency. There are, however, some problems—including some sensitive political ones—with this approach. Increased energy prices inevitably place a greater burden on low income families who devote a significantly higher proportion of their annual budgets to energy expenditures than do wealthy families (Grier 1977). There is also heavy opposition to price increases from consumer groups and a reluctance in Congress to adopt measures that will produce stiff increases in the price of energy. These difficulties are compounded by the uncertainty over the precise effect increased energy prices have on demand. Or as economists phrase it, the "price elasticity of energy demand"—the extent to which a percentage increase in energy prices will correspondingly reduce demand—is not yet fully understood. Concern has thus been expressed that in order to significantly reduce fuel consumption, prices may have to be pushed to highly inflationary levels.

The establishment of mandatory efficiency standards for buildings, appliances, industrial processes, and transportation vehicles offers an alternative method of encouraging efficient energy utilization that could complement the effects of increased prices (Widmer and Gyftopolous 1977) and avoid some of the drawbacks of a sole reliance on increased prices. Standards, in effect, could be set in such a way as to mandate the same degree of efficiency improvements that would be accomplished by allowing energy costs to rise to their "true" replacement value, while avoiding the inflationary impact of elevated prices. The federal government has imposed fuel economy standards on automobiles, but the timetable for these improvements can certainly be stepped up. The Department of Energy is currently drafting energy performance standards for buildings.

Coupon rationing can provide a short-term means of controlling the consumption of oil and natural gas that would avoid some of the discriminatory and inflationary features of "price-induced" rationing (Henderson 1978). However, doubts have been raised over whether such a program could be administered in an intelligent and equable manner. Moreover, rationing is not a suitable tool for controlling consumption over the longer term.

A variety of other measures could be adopted on federal, state, and local levels to eliminate economic and institutional barriers to energy conservation and solar utilization. A critically important level to institute changes is at the research and developmental stage. The current federal program has been heavily slanted towards the development of energy supply technologies, and energy efficiency considerations have been correspondingly neglected. Federal research and development appropriations for all solar technologies over the past decade have amounted to less than 10 percent of the total funding for energy technologies. A fundamental shift in funding priorities must take place in order for solar energy to assume a leading role in the future. In light of their great potential, solar technologies should become the focus of the government's energy research and development program with reducing the cost of harnessing solar energy established as a principal goal. Appropriations should reflect both the developmental status as well as the potential contributions of the individual solar technologies.

Government expenditures in the form of direct subsidies to support public utilization of solar energy, solar demonstration projects, and purchases of solar equipment for government buildings can have a pronounced and positive effect upon a burgeoning solar industry. Government grants can assist low income persons in defraying the cost of solar installations. A well conceived and administered government demonstration program can help gather system performance and reliability data, a process requiring a good deal of time. The sizeable market for solar devices that would be created under a large-scale government procurement program would facilitate the adoption of mass manufacturing techniques and lead to dramatic reductions in the cost of producing solar equipment.

Solar utilization could be spurred by legislation mandating the consideration of installing solar devices in all new and existing government, and perhaps nongovernment, buildings and facilities. For example, a bill sponsored by Senator Paul Tsongas of Massachusetts would require the use of a wide variety of solar equipment on federal buildings and federally funded construction (SEIR 1979b). An ordinance in San Diego, California, will require, effective October 1979, that solar water heating be used in all new residential subdivisions where natural gas is not available. Beginning October 1980, solar heating will be mandated in all new subdivisions even if natural gas is available. The California Energy Commission is exploring the possibility of requiring certain solar applications, primarily domestic hot water and passive heating systems, in all new construction throughout the state where cost effectiveness to consumers can be demonstrated.

Studies have shown that massive government purchases of photovoltaic cells to replace diesel generators at defense installations, for example, could quickly bring the costs of these devices down to competitive levels and save the government money over the long run as well. Legislation passed by Congress to hasten the commercial availability of economic photovoltaic cells by means of federal research, demonstration, and procurement programs is a promising development in this direction. A similarly ambitious federal program is needed to accelerate the development and introduction of wind power systems. It has been suggested that the Department of Energy sponsor a wind power program similar to the government's hydroelectric power development effort, involving the construction of thousands of windmills in the Great Plains to provide electricity and power irrigation pumps throughout the region (CEQ 1978).

A vigorous U.S. program to facilitate the introduction of solar technologies in foreign countries could help alleviate pressure for worldwide fossil and nuclear fuels and also directly help those countries achieve energy and economic self-sufficiency. Many on-site solar devices are likely to be economically attractive in developing countries—having generally higher energy prices and lacking an extensive transmission and distribution system—sooner than in the United States. The expanded market for solar energy equipment opened up by exports abroad could have a beneficial effect on the domestic solar industry and allow for the scaling up of domestic production capability. This could facilitate the implementation of mass production methods by U.S. manufacturers and thereby speed the rate of manufacturing cost reductions, in turn expanding domestic markets and utilization. The increased reliance on solar energy in the United States, coupled with a vigorous solar export business, could improve U.S. trade and balance of payments problems that are largely a result of the present heavy dependence on imported foreign oil.

Investments in conservation and solar use can be significantly encouraged by incentives such as tax credits and low interest loans—measures that in fact formed an important part of the National Energy Act (NEA) passed by Congress in 1978. The NEA specifically offered taxpayers a 30 percent credit on the first $2,000 invested in solar equipment and a 20 percent credit on the next $8,000 spent on qualifying equipment. A 15 percent tax credit was awarded to conservation expenditures of up to $2,000. The state of California offers an even more generous tax credit for solar utilization of 55 percent.

In addition to tax credits, the establishment of alternative financing arrangements can greatly facilitate the introduction of solar technologies. The establishment of an energy development bank—to provide long-term, low interest loans to purchasers of conservation and solar equipment—would be an important first step. A proposal for a Solar Development Bank was included in a set of legislative initiatives sent by President Carter to Congress in June, 1979. However, funding for this bank is contingent upon passage of the controversial "windfall" profits tax.

Utilities also offer significant possibilities for the financing of conservation and solar investments. The widespread utilization of solar energy will have a major impact on utilities. It is therefore important that efforts to introduce energy efficiency measures and solar technologies should be made cooperatively with the utilities, or at least in a non-threatening manner, in order to avoid creating unnecessary barriers and needless delays to large-scale solar implementation (Stobaugh and Yergin 1979).

Utilities have the ability to amass capital in quantities needed to finance the construction of solar energy systems designed on a community (or larger) scale. In many cases, these "district" energy systems may hold economic advantages over smaller, independent systems. Public utilities could ensure that energy is made available to all citizens on a dependable basis and provide a convenient vehicle for subsidizing energy costs for the poor. In this way, the prospect of energy becoming nothing more than a luxury for those wealthy enough to afford their own solar collector panels, photovoltaic arrays, or wind turbines could be avoided.

Utilities may also serve to facilitate consumer investments in energy conservation and solar energy devices for private dwellings. For example, consumers could purchase insulation or solar heating equipment through their local utility, which would bear the responsibility for installation and service. The investment could be gradually repaid in monthly bills, thereby circumventing the problem of the formidably high first cost of solar energy systems and also allaying apprehensions concerning reliability.

An innovative approach to financing energy conservation that can also provide investment opportunities as well as financial benefits to electric utilities is currently underway in the Pacific Northwest (Davenport 1979). Under the so-called "Oregon Plan," the Pacific Power and Light Company, based in Portland, Oregon, has begun to assume responsibility for weatherizing the all-electric homes of its customers (dependent upon owner approval), where conservation measures are deemed cost-effective. The utility company performs a free energy "audit" of a home, and—if it finds that conservation measures such as the installation of additional insulation, weatherstripping, or window glazing will substantially cut down on electricity use—pays for the work to be done. The expense of installing these energy efficiency improvements is entered into the overall utility rate base, thus affording the utility with a return on its investment. The individual customer is not obliged to pay directly for the work done on his or her home unless the house is sold, at which time the "loan" is repaid on an interest-free basis. Pacific Power and Light has found that, in many cases, it is much more economical to reduce electricity demand through the introduction of conservation measures than it is to provide the additional electricity that would otherwise be needed. In fact, the company has found that the cost of electrical power saved through conservation improvements is, on the average, only about half the incremental cost of electricity obtained from the construction of new power plants. This same general financing approach could of course be used by other utility companies and in conjunction with solar energy systems as well.

The government can both educate the public and simultaneously cultivate consumer interest by providing accurate information and dispelling popular misconceptions about solar energy utilization. The government could act as a clearinghouse for current information on the different types of systems available from the various solar manufacturers. Given that the high "first costs" of solar energy systems are undoubtedly the most intimidating consideration from an investor's point of view, the concept of "life cycle costing"—whereby the cost of a solar energy system is compared with cumulative expenditures for conventional energy forms over the expected lifetime of the system—should be widely familiarized.

In order to be able to compete with other energy sources, solar technologies must be made to provide a level of reliability comparable to that currently associated with conventional energy sources. Consumer interests can thus be advanced by a uniform program conducted by the government for the testing and certification of solar equipment. Equipment standards should be designed to afford maximum consumer protection without inhibiting technical innovation in a rapidly growing field. Standards should be flexible in nature, focusing on overall performance rather than specifying design. The government should also take steps to promote competition within the solar industry.

Legislation may also be required on state and local levels to remove any remaining roadblocks to solar utilization. For example, state and local sales taxes and property taxes on solar equipment should be reduced or eliminated to encourage use. In some cases, the revision of local building codes and zoning ordinances may be required to allow for the installation of solar devices. "Sun rights" may also have to be legally established to protect the access to the sun for an investor or potential investor in solar energy systems.

CONCLUSIONS

In the midst of the mounting energy-related difficulties confronting the United States—dwindling and increasingly expensive supplies of oil and natural gas; growing instabilities in the world oil market and the economic and national security problems these bring; problems in the mining, transporting, and burning of coal; and a rapid deterioration of the nuclear power industry—one clear solution emerges: an aggressive strategy emphasizing improvements in energy productivity and the implementation of a variety of attractive solar technologies that can lead us out of the morass and onto the road to a safe and sustainable energy future. We conclude that carefully selected solar energy collection, storage, distribution, and conversion technologies can be integrated into a workable system able to meet the diverse scope of energy needs of the United States in the future. A solar power strategy capable of satisfying these stringent requirements appears to be technically and economically feasible, environmentally and socially desirable, and compatible with healthy economic growth, reduced unemployment, and a high standard of living for all citizens.

A vigorous effort to take advantage of existing and developable cost-effective means of raising the efficiency of energy use, coupled with the rapid introduction of the most promising solar technologies as they become available, can lead to a reduced dependence on imported oil and can also alleviate the severe environmental problems that would necessarily accompany a major expansion of coal, shale oil, and nuclear energy production. The establishment of a renewable energy economy can contribute to a general slowing in the rate of energy-driven inflation—caused by reliance on depletable fuels whose extraction costs rise as the remaining resources become progressively dilute—and an invulnerability to shortages of fuels, either real or contrived. Such an energy system would be free from the impending threat of catastrophic accidents. Environmental degradation could be substantially reduced, and air quality could be dramatically improved, particularly in urban areas. This, combined with the vast potential for employment in solar industries, could lead to the rejuvenation of our urban areas.

These potential benefits can extend far beyond national borders. Internationally, the widespread utilization of solar energy can promote economic growth and industrialization in developing nations, simultaneously diminishing the potential for worldwide conflicts over the supply of oil, reducing the risks of nuclear proliferation, and minimizing the threat of irreversible modification of the environment and global climates.

Changes are certain to occur, whether we like them or not, as our principal energy forms—oil and gas—become increasingly scarce and costly produce. In the absence of concrete alternatives, a heightened dependence on imported oil is an almost inevitable prospect. Strong initiative must therefore be taken to expedite the transition to intelligently selected alternate energy sources. The United States should quickly establish the complementary goals of energy efficiency and solar energy utilization as the centerpiece of national energy policy and move thoughtfully, but rapidly, toward their realization. There is indeed a way out of the present energy dilemma: the solution to the energy "crisis" need not be painful.

REFERENCES

Antal, M.J. 1976. "Tower Power: Producing Fuels from Solar Energy." *Bulletin of the Atomic Scientists* 32, no. 5 (May):59-62.

Behrin, R., et al. 1977. "Energy Storage Systems for Automobile Propulsion, Vol. 1, Overview and Findings." UCRL-52303. Livermore: University of California—Livermore, December.

Bossong, K. 1978. "Resurgence of the Electric Car." Washington, D.C.: Citizen's Energy Project, Report Series no. 36, March.

Chambers, R.S., et al. 1978. "Gasohol: Does It or Doesn't It Produce Net Energy?" Urbana: Energy Research Group, University of Illinois, September.

Chemical and Engineering News (C&EN). 1979a. "Study Finds Chemical Heat Pipe Competitive." 57, no. 5 (January 29):7.

———. 1979b. "Carter Calls Again for Solar Energy Push." 57, no. 26 (June 25):6.

Committee on Nuclear and Alternative Energy Systems (CONAES). 1978. *Report of the Solar Resources Group.* Washington, D.C.: National Academy of Sciences.

Commoner, B. 1979a. "The Solar Transition-1." *The New Yorker*, April 23, pp. 53-98.

———. 1979b. "The Solar Transition-2." *The New Yorker*, April 30, pp. 46-93.

Cotey, V., and M. Dubey, 1976. "The High Potential of Wind as an Energy Source." Second Annual Energy Symposium, Los Angeles Council of Engineers and Scientists, May 19.

Council on Environmental Quality (CEQ). 1978. "Solar Energy: Progress and Promise." Washington, D.C., April.

Davenport, C.P. 1979. "Proposed Residential Energy Efficiency Rider." Proposed Testimony before the Public Utility Commission of the State of Oregon, Exhibit 1-T, Portland Pacific Power and Light Company.

Department of Energy (DOE). 1978. "Status Report on Solar Energy Domestic Policy Review." Washington, D.C., August 25.

———. 1979a. "Modest Impact Seen for Alcohol Fuels Until Conversion Facilities Expand." *Energy Insider* (July 23):1.

———. 1979b. "Response Memorandum to the President of the United States—Domestic Policy Review." TID-22834. Washington, D.C.

———. 1979c. "Energy Supply and Demand in the Midterm." DOE/EIA-0102/52. Washington, D.C.: Energy Information Administration.

Department of the Interior (DOI). 1975. "Energy Perspectives." Washington, D.C.: Government Printing Office, February.

Energy Research and Development Administration (ERDA). 1975. "National Solar Energy Research, Development, and Demonstration Program: Definition Report." ERDA 49. Washington, D.C., June.

Federal Energy Administration (FEA). 1974a. "Project Independence: Solar Task Force Report." Washington, D.C., November.

——. 1974b. "Project Independence: Geothermal Task Force Report." Washington, D.C., November.

Grier, E.S. 1977. "Colder. . .Darker." Prepared by the Washington Center for Metropolitan Studies for the Community Services Administration, Washington, D.C., June.

Henderson, C. 1978. "The Tragic Failure of Energy Planning." *Bulletin of the Atomic Scientists* 34, no. 10 (December):15-19.

Heronemus, W.E. 1973. "Energy Alternatives: The Need and the Possibilities." Lecture Delivered at Hampshire College, November 28.

Inglis, D.R. 1978. "Power from the Ocean Winds." *Environment* 20, no. 8 (October):17-20.

Intertechnology Corporation (ITC). 1977. "An Analysis of the Economic Potential of Solar Thermal Energy to Provide Industrial Process Heat." Warrenton, Va., February.

Karkheck, J., et al. 1977. "Prospects for District Heating in the U.S." *Science* 195 (March 11):948-55.

Kihss, P. 1979. "U.S. Project Seeking to Measure Effectiveness of Electric Vehicles." *New York Times*, February 25.

MacDonald, G.J. "An Overview of the Impact of Carbon Dioxide on Climate." *Bulletin of the American Physical Society* 24, no. 1 (January):31.

McQuiston, J.T. 1979. "Gasohol: The Issue Is Practicality." *New York Times*, May 19.

MITRE Corporation. 1978. "Solar Energy, A Comparative Analysis to the Year 2020." McLean, Va.

National Academy of Sciences (NAS). 1976. "Criteria for Energy Storage R&D." Washington, D.C.

National Science Foundation (NSF)/NASA Solar Energy Panel. 1972. "An Assessment of Solar Energy as a National Energy Resource." College Park: University of Maryland.

New York Times. 1979. "Two Energy-Saving Plans Approved for Postal Vechicles and Offices." May 13.

Office of Technology Assessment (OTA). 1979. "Changes in the Future Use and Characteristics of the Automobile Transportation System—Summary and Findings." Washington, D.C., February.

——. 1979b. "Gasohol: A Technical Memorandum." Washington, D.C.: Government Printing Office, September.

Rattner, S. 1979. "President Setting Solar Power Goal." *New York Times*, June 9.

Rodberg, L.S. 1979. "Employment Impact of the Solar Transition." A study prepared for the Joint Economic Committee, Congress of the United States. Washington, D.C.: Government Printing Office, April 6.

Solar Energy Digest (SED). 1977a. "Billings Starts Converting Datsuns to Run on Hydrogen Fuel." (October):4.

———. 1977b. "Hydrogen Stored in Cylinders Can Now Drive a Car 200 Miles." (February):8.

Solar Energy Intelligence Report (SEIR). 1979a. "Brazilian Ethanol Experience: New Automobile Fuel Comes of Age." 5, no. 16 (April 16):152.

———. 1979b. "Solar Use in Federal Buildings Called for in Bill by Tsongas." 5, no. 15 (April 9):143.

Stanford Research Institute (SRI). 1977. "Solar Energy In America's Future—A Preliminary Assessment." Prepared for the Energy Research and Development Administration. Palo Alto, Calif., March.

———. 1978. "Solar Energy Research and Development: Program Balance." Palo Alto, Calif.

Stobaugh, R., and D. Yergin. 1979. "After the Second Shock: Pragmatic Energy Strategies." *Foreign Affairs* 57, no. 4 (Spring).

Study on Man's Impact on Climate (SMIC). 1971. *Inadvertant Climate Modification*. Cambridge, Mass.: MIT Press.

Widmer, T.F., and E.P. Gyftopolous. 1977. "Energy Conservation and a Healthy Economy." *Technology Review*, June.

Abbreviations and Acronyms

AC: alternating current
AEC: U.S. Atomic Energy Commission
AIA: American Institute of Architects
AIAA: American Institute of Aeronautics and Astronautics
APL: Applied Physics Laboratory
APS: American Physical Society
Btu: British thermal unit
°C: degrees Centigrade
C&EN: Chemical and Engineering News
CEQ: Council on Environmental Quality
cm: centimeter
CONAES: Committee on Nuclear and Alternative Energy Systems
DC: direct current
DOI: U.S. Department of the Interior
DOE: U.S. Department of Energy
EPP: Energy Policy Project of the Ford Foundation
EPRI: Electric Power Research Institute
ERDA: U.S. Energy Research and Development Administration
°F: degrees Fahrenheit
FEA: U.S. Federal Energy Administration
FPC: U.S. Federal Power Commission
ft: foot (feet)
GNP: gross national product
GWe: gigawatt(s) (electrical)
IEA: Institute of Energy Analysis
JPL: Jet Propulsion Laboratory

kV: kilovolt
kW: kilowatt(s)
kWe: kilowatt(s) (electrical)
kWt: kilowatt(s) (thermal)
kWh: kilowatt-hour
kWhe: kilowatt-hour (electrical)
kWht: kilowatt-hour (thermal)
lb: pound
LMFBR: Liquid Metal Fast Breeder Reactor
LNG: Liquefied Natural Gas
LWR: Light Water Reactor
m: meter
MW: megawatt(s)
MWe: megawatt(s) (electrical)
MWt: megawatt(s) (thermal)
NAS: National Academy of Sciences
NASA: National Aeronautics and Space Administration
NRC: U.S. Nuclear Regulatory Commission
NSF: National Science Foundation
OPEC: Organization of Petroleum Exporting Countries
OTA: Office of Technology Assessment
OTEC: ocean thermal energy conversion
SED: Solar Energy Digest
SEIR: Solar Energy Intelligence Report
SRI: Stanford Research Institute
SSPS: solar satellite power station
tcf: trillion cubic feet
UCS: Union of Concerned Scientists
USGS: U.S. Geological Survey
W: watt(s)
W/lb: watts per pound
Wh: watt-hour
Wh/lb: watt-hours per pound

Glossary

active solar energy system—A solar heating and cooling system that relies on an external power source to circulate a working fluid and to distribute heat within a building.

alternating current—An electrical current whose direction reverses at a regular frequency and whose magnitude varies sinusoidally. Electricity in this form is used in the vast majority of applications in the United States.

baseload power plant—An electrical generating facility designed to operate with a relatively constant power output in order to satisfy the maximum continuous level of electrical demand. Large coal and nuclear power plants are typically built for baseload operation.

biomass—A source of solar energy chemically stored in plants and other organic matter as a result of photosynthesis. Biomass energy sources include terrestrial and marine plants and agricultural, forestry, and municipal wastes, all of which can be utilized for their energy content. Biomass may be burned directly or converted into useful fuels such as ethanol, methanol, or methane.

breeder reactor—A nuclear reactor designed to produce not only electricity, but also more fissionable fuels than it consumes. No commercial breeder reactors are currently built or operating in the United States.

British thermal unit (Btu)—The amount of energy required to raise the temperature of 1 pound of water by 1°F. One Btu is equivalent to about 1,055 joules. The nation's annual consumption of energy is currently about 80 quadrillion (or 80×10^{15}) Btu's.

capacity factor—The ratio of the amount of energy generated by a power plant over a given period of time to that which would have been produced if the plant were operating at 100% of its full rated capacity.

cogeneration—The generation of electricity with the utilization of residual heat for industrial process heat, space heating, or other thermal applications. Cogeneration makes very efficient use of primary fuels.

concentrating solar collector—A collector system incorporating the use of reflective surfaces or lenses in order to increase the intensity of solar radiation impinging on a given absorber area.

cryogenics—A branch of physics dealing with very low temperatures. Technological applications include, among others, both superconductivity and the storage of liquid hydrogen.

direct current—An electrical current which flows in one direction only.

efficiency—The ratio of the amount of useful energy delivered by a device to the amount of energy supplied to the device. This is the so-called "first-law" efficiency which is concerned strictly with the quantity of energy involved. The "second-law" efficiency considers not only the gross quantities of energy involved, but also helps keep track of the quality of energy—that is, its availability for doing useful work. It is defined as the ratio of the theoretical minimum amount of energy needed to perform a desired function to the amount of energy actually used by a real device to perform the same function. The second-law efficiency provides a measure of how much a real device or process falls short of that which is theoretically attainable. It is typically less than, sometimes much less than, the efficiency specified by the first law.

electrolysis (of water)—The decomposition of water into gaseous hydrogen and oxygen by the passage of an electrical current through an aqueous solution. It is one of the principal means of producing hydrogen, itself a potentially useful intermediate form of energy.

electromagnetic radiation—Radiation consisting of electromagnetic waves spanning a continuous spectrum of wavelengths including microwaves, radio waves, infrared radiation, visible light, x-rays, and gamma rays.

energy—A quantity that represents the capability of a system to do work. Energy is always conserved; it can neither be created or destroyed, but simply changes its form and quality. Energy exists in a number of different forms all of which theoretically can be converted into one another. Losses inevitably accompany any conversion process. The lost energy is typically in an unusable form, such as very low temperature heat.

ethanol—Ethyl alcohol, a liquid fuel which may be produced from agricultural crops and residues. This is the form of alcohol found in fermented and distilled liquors. It may also be mixed with gasoline to form the mixture "gasohol".

feedstock (petrochemical)—An input of vital chemical substances such as hydrocarbons used in the production of fuels, fertilizers, asphalt, plastics, and a variety of other petrochemical products.

fission—The splitting of a heavy nucleus into two or more parts called fission products (which are usually highly radioactive), accompanied by the release of energy. This process is the energy source of all existing nuclear power plants.

fluidized bed—A combustion system in which a finely crushed, pulverized fuel, typically coal, is suspended in a stream of air, resulting in more complete combustion than in conventional systems.

flywheel—A device that stores rotational kinetic energy in a spinning mass. Flywheels operate in a manner similar to that of a potter's wheel. They have also been installed in subway and trolley systems in the United States to recover energy normally dissipated as heat during braking.

flux—The rate of flow of energy, fluid, or particles. For example, the solar flux is defined as the rate at which solar energy is received over a given surface area.

fossil fuels—Fuels formed from the remains of plant matter over millions of years by extreme pressures exerted from within the earth. Petroleum, natural gas, coal, shale oil, tar sands, and peat are all fossil fuels.

fuel cell (hydrogen/oxygen)—A device that electrochemically combines hydrogen and oxygen to produce electricity and water. A fuel cell performs a process that is precisely the reverse that of the electrolysis of water.

fusion—A process in which certain light atomic nuclei react and combine (i.e., "fuse") to form heavier nuclei occuring with the release of energy. This process is the source of the sun's energy and also that of thermonuclear explosives. Efforts are currently underway to harness it to produce electricity in "controlled" nuclear fusion reactors.

geothermal energy—Thermal or heat energy stored in various forms within the earth. Sources of geothermal energy include underground steam, hot water, hot dry rock, molten rock, and pressurized gases.

gigawatt (GW)—A unit of power equal to one billion watts (see "watt"). Large nuclear power plants in this country typically have a capacity of about one gigawatt. Installed generating capability in the United States is currently over 500 GWe.

gross energy—The total energy used—including all losses, inevitable or avoidable—in performing a given task.

gross national product (GNP)—The total monetary value of all goods and services produced within a country during a single year. The gross national product of the United States is currently about $2 trillion.

head—The elevation difference between two water reservoirs, applicable to hydroelectric power, tidal power, and pumped storage systems, for example. Gravitational potential energy is stored by water held in the upper reservoir.

heat of fusion—The amount of input heat required per unit mass to convert a substance from a solid to liquid form (i.e., "melting"). The same amount of heat is given off when the liquid reverts to solid form (i.e., "freezing").

hydrocarbon—A chemical compound containing only hydrogen and carbon. Oil, natural gas, and coal are all hydrocarbons.

heat pump—A device used in conventional refrigerators which extracts thermal energy from one reservoir and transfers it to another at an elevated temperature. Heat pumps may be used to provide building heating and cooling and industrial process heat. They may be used in conjunction with solar heating and cooling systems.

hydroelectric power—The power of falling water which can be used to drive turbine generators to produce electricity.

hydride—A metallic compound capable of storing hydrogen molecules within its lattice structure.

insolation—Solar radiation received on the earth's surface. Insolation is typically expressed in terms of the amount of energy received per unit surface area over a given period of time. The average rate of solar insolation, averaged over the continental United States and over night, day, and the seasons, is about 60 Btu/square foot/hour or, equivalently, about 18 watts/square foot.

invertor—An electronic device that converts alternating current electricity to direct current or vice versa.

kilowatt (kW)—A unit of power equal to one thousand watts (see "watt").

Kilowatt-hour (kWh)—A unit of energy equal to 3,413 Btu, frequently used in connection with electricity. One kilowatt-hour is the amount of energy produced in an hour by a facility generating power at an average rate of one kilowatt. The United States currently consumes over 2 trillion kWh of electricity a year.

latent heat storage—The storage of thermal energy by means of a phase change in the storage medium (see "phase change").

light water reactor (LWR)—A nuclear reactor that uses normal water as its coolant and moderator and employs slightly enriched uranium as its fuel. There are two commercial types of light water reactors: the boiling water reactor (BWR) and the pressurized water reactor (PWR). All but two of the reactors currently operating in the United States are LWR's.

megawatt (MW)—A unit of power equal to one million watts (see "watt"). Large nuclear and coal-fired power plants are typically rated at about 1,000 MWe.

methane—A colorless, odorless gas that is the main constituent of natural gas. A methane molecule consists of one carbon atom and four hydrogen atoms; it is the simplest hydrocarbon compound.

methanol—Methyl alcohol, commonly referred to as "wood alcohol". It may be produced from wood and other organic waste sources. It is a suitable liquid fuel for transportation vehicles.

mill—A monetary unit equal to one-tenth of a cent frequently used with regard to electricity prices.

net energy—The energy actually used at the point of end use, excluding that previously lost in conversion and delivery.

nonrenewable energy resource—An energy resource that is finite in amount and essentially nonreplenishable. Fossil fuels, nuclear energy, and geothermal energy all represent nonrenewable resources.

ocean thermal energy conversion (OTEC)—An energy conversion system in which the temperature difference between the warm surface water and colder deeper water of tropical oceans is used to drive a turbine generator to produce electricity.

oil shale—A sedimentary rock which contains a solid organic material which can be converted to an oil-like fuel when heated. U.S. reserves of oil shale are immense, amounting to many times the reserves of petroleum. However, it is costly and environmentally damaging to extract.

order of magnitude—Factor of ten; three orders of magnitude is equivalent to a factor of one thousand, and so forth.

passive solar energy system—A solar heating and cooling system consisting of an energy-efficient building designed to utilize natural energy flows to transfer heat inside and outside, as needed, without relying on the forced circulation of a heating or cooling fluid.

peak power—The "peak" load is the maximum demand for power placed on an electrical delivery system. "Peak power" plants are used to cover the wide variations in demands for electricity above a constant level.

For solar energy systems, "peak power" is the power that would be produced by a system when exposed to maximum ("peak") solar radiation of an intensity of about 1 kilowatt/meter^2 or, equivalently, 316 Btu/ft^2-hr. Under these conditions, a solar-electric panel rated at 1 peak kilowatt would generate 1 kilowatt of power. Such a panel, operating with a typical efficiency of 15 percent, would have an area of nearly 7 square meters.

For a wind energy system, "peak" or "rated" power is the constant output power produced when wind velocities equal or exceed a design rated wind velocity, but are below an emergency shutdown velocity. A wind turbine rated at 1 peak kilowatt would generate 1 kilowatt of electricity at the rated wind velocity.

phase change—The physical transformation of a substance from one form (solid, liquid, or gaseous) to another accompanied by the absorption or release of heat.

photosynthesis—The conversion of solar energy by plants and algae into fixed, chemical energy in the form of organic matter.

photovoltaic cell—A semiconductor device, commonly referred to as a solar cell, that directly converts solar energy into electricity.

power—The rate at which work is done or energy delivered, expressed in units of energy per unit time.

power conditioning equipment—Electrical equipment, including invertors and transformers, for altering the voltage, current, and frequency of electricity.

power tower—A specific type of solar thermal conversion system in which solar energy is concentrated onto a tower-top receiver/absorber by an array of tracking mirrors, and the collected thermal energy used to produce electricity in a turbine generator. Its commercial feasibility has not yet been demonstrated.

primary batteries—Electrical batteries which produce electricity until their original chemical materials are consumed. In order to be reused, the electrodes of these batteries have to be replaced. Flashlight batteries are a familiar example of disposable, primary batteries.

primary energy equivalent—A convention adopted to normalize all energy values by expressing all terms on a thermal energy (i.e., "primary energy equivalent") basis. For example, a quantity of electricity or mechanical energy would be expressed in terms of the amount of primary fuel that would be consumed in order to produce an equivalent amount of energy in mechanical or electrical form assuming a 33% conversion efficiency. This number, typical for electrical generation, is used throughout the text.

quad—A unit of energy equal to 1 quadrillion (10^{15}) Btu. Total annual energy use in the United States is slightly under 80 quads at present.

renewable energy sources—Inexhaustible energy sources that are supplied on a continuous or periodically sustained basis. These sources consist of solar energy in its direct and indirect forms including wind, hydroelectric power, biomass, ocean thermal gradients, waves, ocean currents, and tidal power.

reserves—Mineral or fuel deposits of known quantity, quality, and location that can be recovered with current technology at current prices.

resistance (electrical or "ohmic")—A phenomenon analagous to friction which opposes the flow of electrical current in conductors, converting electrical energy into heat. The resistance is equal to the voltage drop along an electrical circuit divided by the current flowing through the circuit. Electric resistance home heaters make use of this process, converting virtually all of the electricity flowing through them into heat.

resources—Mineral or fuel deposits that are either known to exist, but not in such a state that allows for economical recovery using available technology, or those deposits that remain to be identified, but are suspected or probable on the basis of indirect evidence.

salinity gradients—Differences in the salinity between fresh and salt water sources existing in areas where rivers flow into the sea. Salinity gradients represent a potential source of energy that may be harnessed to produce electricity. The practical feasibility of this approach, however, remains to be demonstrated.

secondary battery—An electrochemical device that stores energy in chemical form, although electricity is the input and output form. Unlike primary batteries, these batteries are rechargeable. The lead-acid batteries employed in automobiles are a familiar example of rechargeable, storage batteries.

semiconductor—A class of materials, including silicon and germanium, whose electrical conductivities lie between those of metal conductors and insulators. Semiconductors are essential materials for transistors and solar photovoltaic cells.

sensible heat storage—The storage of thermal energy by raising the temperature of the storage medium. Most storage systems of this type employ either water, rock, or earth as the storage medium.

solar collector—A device designed to capture solar energy in the form of heat. It may simply consist of a blackened metal sheet covered with glass (i.e., a "flat-plate" collector) or it may use a complex array of lenses or mirrors to focus solar radiation. Collectors may either be stationary or they may be steerable in order to follow ("track") the sun.

solar radiation—Radiant energy received from the sun both directly and indirectly from scattering and reflection.

specific energy—The ratio of energy stored in a device to its mass. This term is used in connection with energy storage devices such as batteries or flywheels. The specific energy of a battery, for example, is the critical determinant of the range of an electric vehicle: the higher the specific energy, the greater the range.

specific heat—The heat necessary to raise the temperature of a unit mass of a specific material by one degree. The specific heat of water is 1.0 Btu/lb/°F, while that of rocks is typically about 0.20 Btu/lb/°F.

specific power—The ratio of the power output of a device to its mass. This term is used in connection with energy storage devices such as batteries or flywheels. The specific power of a battery, for example, is the critical determinant of the acceleration of an electric vehicle: the higher the specific power, the better the acceleration.

subsidence—The collapse of a land surface as a result of the previous collapse of underlying mined areas.

superconductivity—A property of certain metallic conductors which lose all electrical resistance when their temperatures are reduced below a critical "transition" temperature which varies depending upon the material. Such temperatures are typically near absolute zero (minus 460°F) and are therefore difficult, as well as expensive, to achieve.

total energy system—A system designed to provide both useable thermal and electrical energy in a manner that minimizes energy lost in the form of waste heat.

transformer—A device for changing the voltage and current of alternating current electrical energy.

voltage—Electric potential difference capable of inducing an electrical current between two points, analagous to pressure causing water to flow in a pipe. One volt is equal to the potential difference that will cause a current of one ampere to flow through a conductor with a resistance of one ohm.

waste heat—Low-grade thermal energy given off in the generation of electricity which is typically discharged to the environment.

work—The application of a force through a distance. Work is done, for example, when a moving car is pushed, but not when one pushes against a stationary wall. Work is expressed in the same physical units as energy, which in turn is defined as the ability to do work.

Units and Conversion Factors

ABBREVIATIONS

Btu: British thermal unit
cal: calorie
cm: centimeter
eV: electron volt
ft: foot (feet)
hr: hour
J: joule
kg: kilogram
km: kilometer
kWh: kilowatt-hour
m: meter
mm: millimeter
psi: pounds per square inch
W: watt
Wh: watt-hour

PREFIX ABBREVIATIONS

G	giga	10^9
M	mega	10^6
k	kilo	10^3
m	milli	10^{-3}
μ	micro	10^{-6}
n	nano	10^{-9}

ENERGY EQUIVALENTS

Crude Oil
one barrel (42 gallons):   5.6 million Btu
one million barrels per day: 2.1 quads per year

Natural Gas
one standard cubic foot: 1,035 Btu

Coal
one ton: 25 million Btu

Uranium
one pound (completely fissioned): 34 billion Btu

Biomass
one dry ton: 15 million Btu

One quad (primary energy equivalent):
eleven 1,000 MWe power plants operated continuously for one year with a 100% capacity factor.
seventeen 1,000 MWe power plants operated for one year with a 65% capacity factor.

CONVERSION FACTORS

Length:
1 foot (ft) = 0.305 m
1 mile = 5,280 ft
      = 1.609 km

Mass:
1 pound (lb) = 0.454 kg
1 ton (short) = 2,000 lb
            =   907 kg
            = 0.907 tons (metric)

Time:
1 hour = 3,600 seconds
1 year = 8,760 hours

Energy:

$$1 \text{ Btu} = 1{,}055 \text{ J}$$
$$= 2.93 \times 10^{-4} \text{ kWh}$$
$$= 252 \text{ cal}$$
$$= 6.59 \times 10^{21} \text{ eV}$$
$$= 777.5 \text{ foot-pounds}$$

Energy Density (energy per unit area):

$$1 \text{ Btu/ft}^2 = 1.136 \text{ J/cm}^2$$
$$= 3.16 \text{ Wh/m}^2$$
$$= 0.271 \text{ cal/cm}^2$$

Power (energy per unit time):

$$1 \text{ Watt (W)} = 1 \text{ J/sec}$$
$$= 3.413 \text{ Btu/hr}$$
$$= 0.239 \text{ cal/sec}$$
$$= 1.34 \times 10^{-3} \text{ horsepower}$$

Flux (energy per unit area per unit time):

$$1 \text{ Btu/ft}^2/\text{hr} = 3.16 \text{ W/m}^2$$
$$= 0.271 \text{ cal/cm}^2/\text{hr}$$

Area:

$$1 \text{ ft}^2 = 9.3 \times 10^{-2} \text{ m}^2$$
$$1 \text{ mile}^2 = 27.88 \times 10^6 \text{ ft}^2$$
$$= 640 \text{ acres}$$
$$= 2.59 \text{ km}^2$$

Volume:

$$1 \text{ ft}^3 = 28.317 \times 10^{-3} \text{ m}^3$$
$$= 28.317 \text{ liters}$$
$$= 7.48 \text{ gallons (U.S.)}$$
$$= 2.296 \times 10^{-5} \text{ acre-feet}$$

Pressure:

$$1 \text{ psi} = 6{,}895 \text{ newtons/m}^2$$
$$= 6.805 \times 10^{-2} \text{ atmospheres}$$
$$= 51.7 \text{ mm of mercury (at } 32°\text{F)}$$

Temperature:

$$°F = (1.8 \times °C) + 32$$
$$°K \text{ (Kelvin)} = °C + 273.15$$

Index